MOBILE CHANNEL
CHARACTERISTICS

THE KLUWER INTERNATIONAL SERIES
IN ENGINEERING AND COMPUTER SCIENCE

MOBILE CHANNEL CHARACTERISTICS

by

James K. Cavers
Simon Fraser University,
Canada

Springer Science+Business Media, LLC

Electronic Services <http://www.wkap.nl>

Library of Congress Cataloging-in-Publication Data

Cavers, James K., 1944-
 Mobile channel characteristics / by James Cavers.
  p.cm.-- (The Kluwer international series in engineering and computer science)
 Includes bibliographical references and index.

Additional material to this book can be downloaded from http://extras.springer.com.
 ISBN 978-1-4757-8417-6   ISBN 978-0-306-47032-5 (eBook)
 DOI 10.1007/978-0-306-47032-5

 1. Mobile communication systems. 2. Telephone communication channels. 3.
Radio--Transmitters and transmission--Fading. I. Title. II. Series.

TK6570.M6C38 2000
621.3845--dc21

                                                                    00-056134

Printed on acid-free paper.

To Penny

ACKNOWLEDGEMENTS

Talk to anyone who has written a book. They will tell you that they wouldn't have believed at the start how much time it would take. They will also tell you that they were helped by many other people who contributed advice or effort. I have been particularly fortunate in this respect, and would like send special thanks to:

- My son, Stephen Cavers, for his expert and calming advice on Windows and computers, generally. He also prepared the graphics used in the installation, including the elegant cell phone icon that should be sitting on your computer's desktop right now.

- Dave Michelson, for sharing his encyclopedic knowledge of propagation phenomena and giving me tips regarding wonderful websites. Any errors in the text are my unique contribution, not his.

- Lisa Welburn and Steve Grant, for their preparation of the Index of Topics - a mind-numbing but essential piece of business, and one I am happy not to have had to do myself.

- Jennifer Evans, of Kluwer Academic Publishers, for sharing my enthusiasm about a hyperlinked, web-savvy interactive text, and for her guidance in shaping it to a publishable form.

TABLE OF CONTENTS

1. Path Loss   1
   1.1 Plane earth model and inverse fourth power   2
   1.2 Implications for cellular design   5
   1.1A Detailed derivation of inverse fourth power   11

2. Shadowing   15
   2.1 Statistical model of shadowing   16
   2.2 Shadowing and system design   19

3. Fading and Delay Spread   23
   3.1 Physical basis of fading   25
   3.2 Mathematical model of fading   28
   3.3 Consequences: Doppler spread, delay spread and time-variant filtering   31
   3.4 Is it Flat or Frequency Selective? An Example   36

4. First Order Statistics of Fading   39
   4.1 Gaussian models in time and frequency   41
   4.2 Rayleigh and Rice fading   44
   4.3 Consequences for BER   51
   4.4 Connecting Fading, Shadowing and Path Loss   54

5. Second Order Statistics of Fading   59
   5.1 Doppler spectrum and WSS channels   61
   5.2 Power delay profile and US channels   69
   5.3 Scattering function and WSSUS channels   77
   5.4 A TDL channel model for analysis and simulation   84

6. Connecting Fading Statistics With Performance   91
   6.1 Random FM and error floor   92
   6.2 Level crossing rate and error bursts   98

7. A Gallery of Channels   105
   7.1 Macrocells   106
   7.2 Urban microcells   110
   7.3 Indoor picocells   113

8. Differences between mobile and base station correlations   115
   8.1 Directionality at the mobile - Doppler spread   116
   8.2 Directionality at the mobile - delay spread   124
   8.3 Angular dispersion and directionality at the base   126

9. Simulating Fading Channels   137
   9.1 Sampling and SNR   138
   9.2 Complex gain generation   145
   9.3 Importance sampling   148
   9.3A Effect of β on variance in importance sampling   157

x

Appendix A: The Lognormal Distribution 161

Appendix B: A Jakes-Like Method of Complex Gain Generation 165

Appendix C: Visualization of a Random Standing Wave 175

Appendix D: Animation of Complex Gain, Fading and Random FM 177

Appendix E: Animation of Time Varying Frequency and Impulse Responses 181

Appendix F: Joint Second Order Statistics 185

Appendix G: Joint Probability Density Functions in Polar Coordinates 191

Appendix H: Generation of Complex Gain for Importance Sampling 197

Appendix I: Linear Time Variant Filters 201

Appendix J: Is the Channel Really Gaussian? An Experiment. 205

Appendix K: The FFT and Overlap-Add Method of Complex Gain Generation 209

References 217

Index of Topics 223

PREFACE

THIS IS NOT A BOOK.

IT IS A JUST A PAPER COPY OF THE CD-ROM CONTENTS.

This is Not a Conventional Book

Mobile Channel Characteristics was conceived and written as an interactive text to be viewed on a computer screen. It includes many features not found in conventional texts:

- The entire text resides on your hard drive. It appears as an icon on your desktop, and it's always ready, just a mouse click away.
- It is a live document. Try different parameter values, and the equations, tables and graphs recalculate as you watch. Animated graphs illustrate dynamics of the channel. Explore propagation, modulation or system models interactively to gain additional insight.
- The examples and appendices are "tear-off design sheets". Use their programs on the job or in your thesis to speed up your work - examples include three good methods for generating channel gain process for your simulations.
- It links you to the world. Hyperlinks connect you to websites of cited authors, to online research journals and to employers and graduate schools.

Scope and Objectives

This text introduces models of the major phenomena of mobile communications: path loss, shadowing and the delay spread and Doppler spread introduced by multipath scattering. You need to understand them before you study more advanced topics, some of which are covered in other texts in this series: *Detection and Diversity*, *Coding for Mobile Communications* and *Smart Antennas*.

The presentation is designed to be accessible to senior undergraduates or first-level graduate students. In scope, it is about one third of a standard one-semester course. That is, this material would normally be covered in about 13 lecture hours. Add on two hours of study for each hour of lecture and you have almost 40 hours. That's roughly the time you can expect to spend in self-study to master the topics.

As for objectives, there are really just two:
- to help you build intuition and a quick understanding of the major phenomena;
- and to provide you with some computational tools for work in the area.

Real understanding requires a lot of math - but simply dropping you into the deep end of a pool of equations won't help much. Instead, I've used hyperlinks to construct the course on several levels of detail. The top level is a general discussion. If you're not interested in a more mathematical treatment, it is sufficient for basic understanding.. If you do want to follow the arguments in more detail, though, you can use the hyperlinks and Appendices to go deeper.

About the CD-ROM

Mobile Channel Characteristics was conceived and written as an interactive text to be viewed on a computer screen. It includes many features not found in conventional texts:

- The entire text resides on your hard drive. It is always ready, just a mouse-click away.
- It is a live document. Try different parameter values, and the equations, tables and graphs recalculate as you watch. Animated graphs illustrate dynamics of the channel. Explore propagation, modulation or system models interactively to gain additional insight.
- The examples and appendices are "tear-off design sheets". You can use their programs on the job or in your thesis to speed up your work.
- It links you to the world. Hyperlinks connect you to websites of cited authors, to online research journals and to employers and graduate schools, all through the Internet.

Installing with Windows 95/98 and Windows NT 4.0

Insert the CD-ROM in your CD-ROM drive. The installation dialog begins automatically and you will be guided through the installation. *Mobile Channel Characteristics* occupies 9.5 Mbyte of hard drive space.

You must have version 8 or higher of Mathsoft's Mathcad or Mathcad Explorer on your computer **at the time you install this text**. If you do not have either of them, the installation dialog helps you download a free copy of Mathcad Explorer from Mathsoft (www.mathsoft.com), **provided you are connected to the Internet.** The download file is 12.4 Mbyte. It is recommended that you run Mathcad Explorer once immediately after its installation before the first time you open *Mobile Channel Characteristics*.

One section of *Mobile Channel Characteristics* contains images which can be viewed more clearly with Adobe Acrobat Reader. This is optional, not essential. If you do not have it, the installation dialog allows you to install Adobe Acrobat Reader 4.05 directly from the CD-ROM.

Disclaimers

This CD-ROM is distributed by Kluwer Academic Publishers with *ABSOLUTELY NO SUPPORT* and * NO WARRANTY * from Kluwer Academic Publishers.

Kluwer Academic Publishers shall not be liable for damages in connection with, or arising out of, the furnishing, performance or use of this CD-ROM.

1. PATH LOSS

At the macroscopic level of system layout, the most important issue is path loss. In the older mobile radio systems that are limited by receiver noise, path loss determines SNR and the maximum coverage area. In cellular systems, where the limiting factor is cochannel interference, path loss determines the degree to which transmitters in different cells interfere with each other, and therefore the minimum separation before channels can be reused.

Accurate prediction of path loss in a specific geographic setting is a challenging task. In fact, many consultants and large service providers maintain proprietary software packages to do just that during the initial layout phase or subsequent redesign of cellular systems. These packages account for the physical topography, the building height and density and the anticipated traffic density, and make use of a variety of models to account for reflections, diffraction and distance effects.

This text, however, does not treat the subject in detail. Instead, we will use simplified models that apply to generic situations in which only the distance from the base antenna is of importance. Variability caused by local conditions - shadowing - is considered separately in Section 2. Despite their simplicity, though, the models are sufficient to explain many issues in cellular design, and are used by most authors as the first approximation when introducing new system concepts.

Virtually all of the standard references, for example [**Stub 96, Stee92, Yaco93, Jake74**], deal with path loss. These notes provide a summary and some different perspectives.

1.1 Plane earth model and inverse fourth power

This section uses a very simple model to establish:

* the plausibility of inverse fourth power path loss asymptotically;
* the reason for large variations of path loss close to the base antenna;
* the existence of two regimes: slow decrease with distance near the base, and faster decrease with distance after some break point.

1.2 Implications for cellular design

The inverse fourth power path loss - instead of the more familiar inverse square law - has important consequences for system design. Some of them are unfortunate, like the loss of SNR, but others work in our favour, such as the reduction of cochannel interference into other cells.

1.1A Detailed Derivation of Inverse Fourth Power With Two Rays

This chapter appendix simply provides more detail for the discussion in Section 1.1 on the asymptotic inverse fourth power behavior.

1.1 Plane earth model and inverse fourth power

The familiar inverse square law of free space propagation is far too optimistic in mobile communications, where the path loss exponent ranges typically between 3 and 4. Why? We'll see below that it results from interference between grazing reflections from large geometry objects.

The plane earth is the simplest large geometry object. In real life (other than, say, the North American prairies or the Australian outback), the situation is much more complex. Nevertheless, we can learn a lot from this simple model. Consider the sketch below, with a direct and a reflected ray (i.e., a normal to a plane wave front) between base and mobile, and an assumed reflection coefficient of -1. The base station and mobile antennas are at heights h_b and h_m and are separated by a horizontal distance d. Note that the size, or aperture, of the antennas is a separate issue.

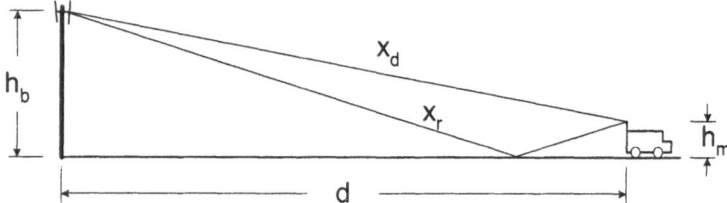

Interference between the two rays causes all the interesting action. Interference depends on phase difference, which in turn depends on path length difference, which is the quantity we'll now examine. The path length difference is greatest, at $2h_m$, when the mobile is right beside the base antenna ($d=0$). As the mobile moves away from the base, the path length difference decreases, approaching zero for very large distances. From this, and the fact that the reflection coefficient is very close to -1, so the paths subtract, we can deduce most of the behaviour:

* There are several oscillations in signal strength as the mobile moves out, since the paths alternately cancel or reinforce. How many oscillations? Roughly $2h_m/\lambda$, the number of wavelengths in the initial path length difference.

* After the path length difference has decreased to $\lambda/4$, the rest of the distance out to infinity (or at least Chicago) is characterized by a slow decrease of phase difference from $\pi/2$ down to zero.

* In that outer region where phase difference is less than $\pi/2$, we observe inverse fourth power: an inverse square law from the natural propagation decrease of each ray by itself, and a second inverse square law from the rate of decrease of phase difference, arising from the geometry. The product of the two phenomena is inverse fourth power. The **detailed derivation** makes use of straightforward geometry, followed by some series approximations in the distant inverse fourth power region.

Now let's look at the behaviour graphically. The power gain function is derived in the detailed section, and is made available here through a file reference.

⊡ Reference:D:\COURSES\MobChann\paperbook\Turaydet.mcd(R)

⊡ Reference:D:\COURSES\MobChann\paperbook\Units.mcd(R)

Try experimenting with different antenna heights *expressed in wavelengths*:

$h_b := 30$ $\qquad h_m := 3$

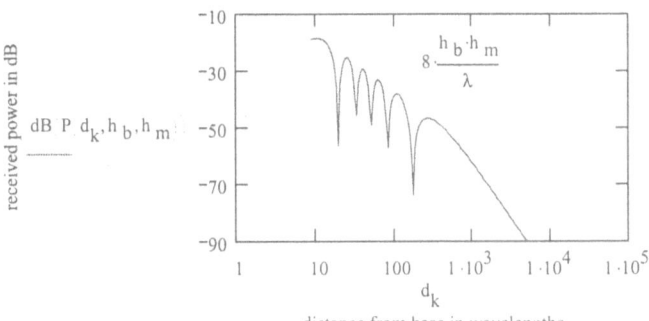

distance from base in wavelengths

You can see that there are two regimes: near the base, there are strong ripples and the maxima drop according to inverse square law; and past the last maximum, the signal drops steadily as an inverse fourth power. You should also have found that there are twice as many maxima as there are wavelengths in the mobile antenna height. This is not surprising, since the differential path length decreases from $2h_m$, when the mobile is directly under the base antenna, to almost zero, when the mobile is far away from the base. If there are more than two rays, then there are more maxima.

As noted earlier, the inverse fourth power regime is due to a combination of natural inverse square law and the increasing degree of cancellation between the two rays as the differential path length approaches zero. The series approximations in the **detailed derivation** show that the received power is approximately

$$\frac{1}{2} \cdot \frac{1}{d^2} \cdot P_s \cdot \left(\frac{4 \cdot \pi \cdot h_b \cdot h_m}{\lambda \cdot d} \right)^2 \qquad \text{which varies as} \qquad \frac{h_b^2 \cdot h_m^2}{d^4} \qquad (1.1.1)$$

where P_s is the transmitted power. Therefore the received power is inversely proportional to d^4, and is directly proportional to the squared antenna heights. The dependence on wavelength λ is more complicated than shown here, because atmospheric attenuation varies significantly with wavelength, as do the antenna power gains (for horn or planar antennas, the gains are inversely proportional to λ^2). In any case, the essence of the model is a received power

$$P = \frac{P_o}{d^4} \qquad (1.1.2)$$

where P_o is a proportionality constant.

The break distance that separates the two regimes is (see the graph)

$$8 \cdot \frac{h_b \cdot h_m}{\lambda} \qquad (1.1.3)$$

This is just the distance at which the phase difference between the two rays is down to $\pi/2$. The final maximum is at half that distance.

Even though we used a grossly simplified model, it is interesting that measured power in an urban microcell [**Gree90**] behaves much like the graph above, although the presence of more than two rays produces more maxima in the inverse square law regime. Microcell measurement: reported in [**Taga99**], shown below, also demonstrate distinct inverse square and inverse fourth power regimes with a break point at about 150 metres.

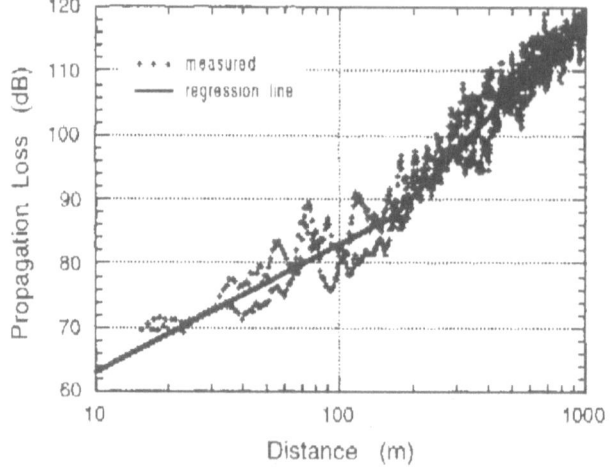

Click on this thumbnail for a clearer picture.

Propagation Loss Characteristics of a Typical Urban Road
from [**Taga99**] (© 1999 IEEE)

As for our inverse fourth power regime, measurements of path loss in urban and suburban macrocell environments [**Hata80**] show that the exponent generally varies between 3 and 4, and is usually closer to 4. In urban microcells, the exponent ranges from about 4 to over 9 [**Gree90**]. A recent examination of the propagation model can be found in [**Benn96**]. For a good survey of path loss mechanisms and models, see [**Yaco93**]. Discussion of various measurement studies and their limits of applicability is contained in [**Stee92**].

1.2 Implications for cellular design

The inverse fourth power path loss - instead of the more familiar inverse square law - has important consequences for system design. Some of them are unfortunate, but others work in our favour. When analysing the effect of propagation loss, we'll have to distinguish between two general cases:

* In *noise-limited operation*, we try to maximise the coverage area of a single transmitter, or try to minimise the transmitter power required to give a reasonable noise margin at the outer limits of the coverage area. This is typical of dispatch systems, paging systems or pre-cellular mobile phone systems. Here inverse fourth power is a headache.

* In *interference-limited operation*, the problem is cochannel interference from other cells, more than receiver noise. This is typical of all cellular systems, ones that partition a geographic region into cells, many of which use the same set of time and/or frequency slots. In the case of CDMA, all cells use the same frequencies. Here, inverse fourth power actually helps the system layout.

Noise Limited Operation

In noise limited operation, there is no shortage of problems introduced by by inverse fourth power propagation, compared to inverse square law. **First**, it is obvious that we need a more powerful transmitter in order to achieve reasonable range. Unfortunately, the problem is even worse than that: an increase in transmit power does not give a proportional increase in coverage area. To see this, consider a circular coverage region with radius R_1, base station power P_1 and inverse kth power propagation loss. That is, the power at any distance r is proportional to

$$\text{power} = \frac{P_1}{r^k} \qquad (1.2.1)$$

The edge of the coverage area (the cell radius) is defined by a specific minimum power level (actually, a minimum SNR requirement): $P_{min} = P_1/R_1^k$. If we change the transmitter power to P_2, the minimum power requirement results in a new allowable radius, according to

$$P_{min} = \frac{P_1}{R_1^k} = \frac{P_2}{R_2^k} \qquad (1.2.2)$$

and the ratio of cell areas is

$$\frac{A_2}{A_1} = \left(\frac{R_2}{R_1}\right)^2 = \left(\frac{P_2}{P_1}\right)^{\frac{2}{k}}$$

If $k=4$, then we need a fourfold (6 dB) increase in power to double the coverage area. If $k=2$ (inverse square law) then power and area are proportional. $\qquad (1.2.3)$

A second consequence of the rapid drop in SNR with distance from the base station shows up when we consider how the users are distributed within the coverage area. If they are scattered uniformly, then we expect more of them at larger distances, since the area of any annulus of thickness Δr is proportional to r. That means that a large fraction of the population is near the edge of a cell and experiences SNR close to the minimum!

Let's examine this issue quantitatively. What fraction of the user population has a given receive power (or SNR)? We can use this function in calculations of outage or user-ensemble error rates. To start, define a circular coverage area of radius R and a received power of at least P_{min}. If we assume that users are uniformly distributed in the service area, then the probability that a user is at at distance r or greater is

$$\Pr(\text{dist} \geq r) = \frac{\pi \cdot R^2 - \pi \cdot r^2}{\pi \cdot R^2} = 1 - \left(\frac{r}{R}\right)^2 \tag{1.2.4}$$

From (1.2.2), the distance at which a user receives power S is

$$r = R \cdot \left(\frac{P_{min}}{S}\right)^{\frac{1}{k}} \tag{1.2.5}$$

Combining (1.2.4) and (1.2.5), the probability that the received power P is less than S is

$$\Pr(P \leq S) = \Pr\left[\text{dist} \geq R \cdot \left(\frac{P_{min}}{S}\right)^{\frac{1}{k}}\right] = 1 - \left(\frac{P_{min}}{S}\right)^{\frac{2}{k}} \tag{1.2.6}$$

For convenience, define the normalized power $U = S/P_{min}$ to obtain the cumulative distribution function

$$F_U(u, k) := 1 - \left(\frac{1}{u}\right)^{\frac{2}{k}} \qquad u \geq 1 \tag{1.2.7}$$

This is the fraction of the population that has effective SNR less than or equal to $u\,SNR_{min}$ if the propagation is inverse kth power.

We'll have a look at it on the next page. First, the plot parameters:

$$i := 0..20 \qquad a := 10^{0.1} \qquad u_i := a^i$$

⬛ Reference:D:\COURSES\MobChann\paperbook\Units.mcd(R)

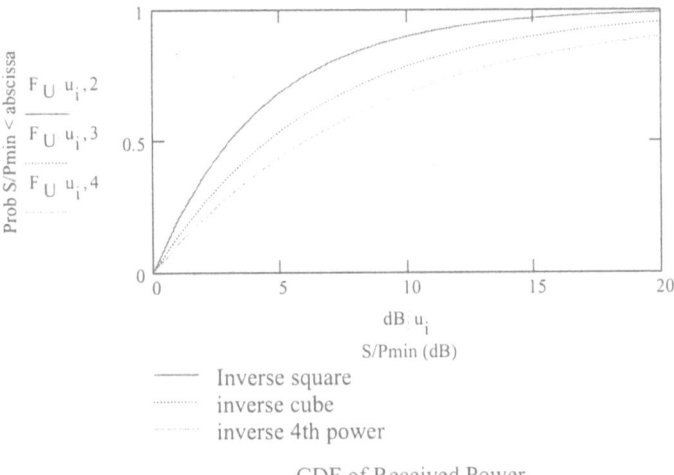

CDF of Received Power

This is an eye-opener! With inverse square law, half of the users have SNR no better than 3 dB above the worst case at the edge of the cell. Inverse 4th power is better, because signal strength climbs faster as we move toward the base, but we still have 30% of the users within 3 dB of the worst case SNR.

Because the CDF in the graph is close to linear for small values of signal power in dB, we can obtain a quick rule of thumb. Define $u=\exp(0.23v)$, where v is the dB equivalent of u. Substitute in the cdf (1.2.7), and expand in a series for small values of v to obtain

$$F_U(v) = \frac{0.46}{k} \cdot v \qquad 0 \cdot dB \le v < 3 \cdot dB \qquad \text{(approx)} \tag{1.2.8}$$

which says that roughly **a fraction 1.5/k of the users have power within 3 dB of the minimum**

A third problem due to propagation loss is termed the "near-far effect". Consider the uplink; that is, reception by the base station of signals from mobiles at various locations in the coverage area. Nonlinearities in the radio transmitters can create distortion that spills into neighbouring channels. If the signals are of comparable strength, this adjacent channel interference (ACI) is usually small enough not to be a problem. However, if the interference is from a nearby mobile, with a strong signal, and the desired signal is from a distant mobile with a signal tens of dB weaker, we have the near/far problem. It is accentuated in inverse 4th power conditions, since signal strength drops rapidly as the mobile moves far away from the base. CDMA systems, in which all signals occupy the same frequency (but with processing gain for some protection), would be fatally compromised by the near far problem. The solution in most cases is to incorporate automatic changes in transmit power as the distance changes ("dynamic power control") to keep the received power roughly constant, and the problem disappears.

Interference Limited Operation

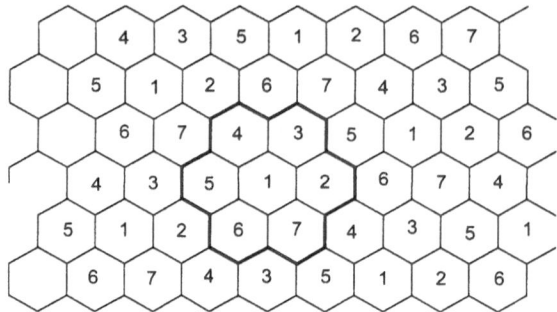

Cellular systems are limited more by interference from other cells than by noise. The sketch above shows an idealised hexagonal array of cells, in which all cells with the same label ("cochannel cells") use the same time/frequency slots and therefore interfere with each other. In this sketch, there are 7 different labels in a "cluster" (one such cluster is indicated by heavy lines). If the level of cochannel interference - transmissions in one cell interfering with reception in another - is not acceptable, the cochannel cells must be separated further. This requires the number of cells in a cluster (the "cluster size") to increase - perhaps to 12, or even to 20. It's a simple solution, but there is a price to be paid: in the sketch, only 1/7 of the total system bandwidth resource (e.g., number of channels) is available in each cell to handle the traffic load. If the cluster size is larger, there is a further dilution of system resource within each cell and a consequent decreased ability to handle the traffic in each cell. Traffic issues are one of the topics in the text *Cellular System Design* in this series.

In CDMA systems, all cells are cochannel (a cluster size of 1). The cochannel interference (CCI) is mitigated by the processing gain of the signals, and by the propagation loss experienced by the adjacent, as well as distant, cells.

In cellular arrangEments, the propagation law plays a critical role. Ideally, we would see zero loss up to the edge of the cell, and infinite loss beyond that distance. We can't have that, of course, but cell boundaries are more sharply defined with fourth power than with second power. The sketch below illustrates the point for inverse square, cube and fourth power laws, assuming that we have been able to meet the minimum power requirement P_{min} at the cell boundary R.

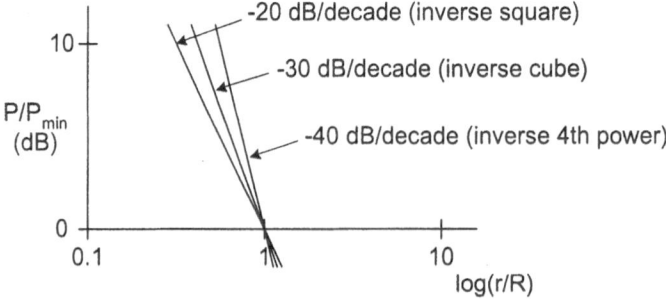

In fact, some urban microcells are even better. As we saw above, the two ray model has a relatively slow inverse square law propagation (albeit with some nulls) out to a break point, after which it drops as inverse fourth power. From the graph, that break point is between 4 and 8 times $h_b h_m / \lambda$ (see (1.1.1) in **Section 1.1**). If the mobile, or handheld unit, has an antenna 5 wavelengths above the ground, then a 5 meter high base antenna puts the break point at about 150 m - just about right for a microcell.

Let's see how propagation exponent, required carrier to interference ratio C/I and cluster size play against each other. An approximate analysis reveals the main features without much work. Consider the case of a mobile at the edge of its cell (as with noise-limited operation, this is the worst location), trying to listen to its centrally-located base station in the presence of interference from base stations in other cochannel cells. The situation is sketched below for a cluster size of 3. Cells have radius R and cochannel bases are separated by D.

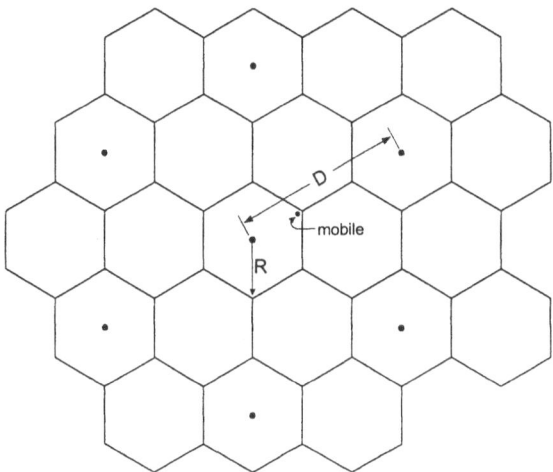

Here's a rough analysis (watch for the word "roughly"). The signal powers received from the desired base and the closest cochannel base are

$$P_d = \frac{P_o}{R^k} \qquad P_i = \frac{P_o}{(D-R)^k} \qquad (1.2.9)$$

Pessimistically, we approximate all 6 nearest interfering bases as having the same power as the closest one (this roughly compensates for the fact that we'll ignore more distant interfering bases). The carrier to total interference ratio is then roughly

$$CIR_{min} = \frac{P_d}{6 \cdot P_i} = \frac{1}{6} \cdot \left(\frac{D-R}{R}\right)^k = \frac{1}{6} \cdot \left(\frac{D}{R} - 1\right)^k \qquad (1.2.10)$$

Next, the cluster size is roughly the ratio of the area of the circle of radius D to the amount of its area taken up by the cochannel cells. Roughly half of each of the 6 cells of the first cochannel tier lies in the circle. Therefore we have a cluster size of about

$$N = \frac{2 \cdot \pi \cdot D^2}{\left(1 + \frac{6}{2}\right) \cdot 2 \cdot \pi \cdot R^2} = \frac{1}{4} \cdot \left(\frac{D}{R}\right)^2 \tag{1.2.11}$$

Combining (1.2.10) and (1.2.11), we have

$$CIR_{min} = \frac{1}{6} \cdot \left(2 \cdot \sqrt{N} - 1\right)^k \tag{1.2.12}$$

This shows clearly that we can achieve a given CIR_{min} with a smaller cluster size, and therefore less dilution of system resources, if the propagation exponent is larger. Let's have a look graphically.

$$i := 0..8 \qquad CIRdB_i := 12 + i \qquad CIR_i := nat(CIRdB_i)$$

$$cluster(CIR, k) := \left[\frac{(6 \cdot CIR)^{\frac{1}{k}} + 1}{2}\right]^2$$

This inverts (1.2.12) to give N in terms of CIR_{min}. (1.2.13)

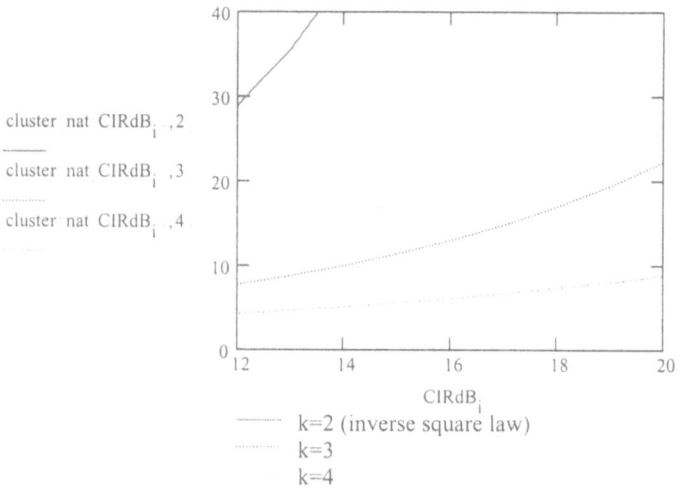

cluster nat $CIRdB_i$, 2

cluster nat $CIRdB_i$, 3

cluster nat $CIRdB_i$, 4

$CIRdB_i$

——— k=2 (inverse square law)
.... k=3
k=4

Required Cluster Size vs. CIRmin

The cluster size N is smaller and increases more slowly with protection ratio (CIR_{min}) if the propagation exponent is higher - and the smaller cluster size gives a more efficient system because each cell gets more of the system resources.

1.1A Detailed Derivation of Inverse Fourth Power With Two Rays

To make the derivation clearer, the sketch is redrawn below to show that the reflected ray can also be considered a direct ray, but from a reflected base antenna.

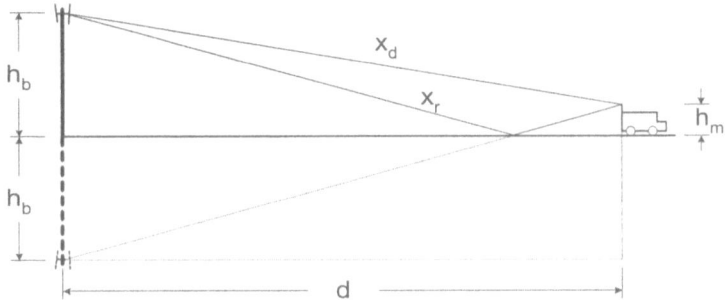

As a first step, we'll get the exact expression for the complex gain and the power gain of the two ray transmission path. The complex envelope at a distance x along each ray is proportional to

$$s\left(t - \frac{x}{c}\right) \cdot \frac{\exp(-j \cdot \beta \cdot x)}{d} \tag{1.1A.1}$$

where $s(t)$ is the transmitted complex envelope due to the modulation and $\beta = 2\pi/\lambda$ is the wave number, where λ is the wavelength. The signal amplitude depends inversely on d because of inverse square law. Because the reflection coefficient is -1, the sum of the arriving rays is

$$\text{sum}(t) = s\left(t - \frac{x_d}{c}\right) \cdot \frac{\exp(-j \cdot \beta \cdot x_d)}{d} - s\left(t - \frac{x_r}{c}\right) \cdot \frac{\exp(-j \cdot \beta \cdot x_r)}{d} \tag{1.1A.2}$$

$$\blacksquare = s\left(t - \frac{d}{c}\right) \cdot \exp(-j \cdot \beta \cdot d) \cdot \frac{1 - \exp(-j \cdot \beta \cdot \Delta x)}{d} \qquad \text{(approx)} \tag{1.1A.3}$$

This approximation is based on narrowband modulation, in which the time scale is much larger than the differential delay $\Delta x/c$, so that $s(t - x_d/c) \approx s(t - x_r/c)$, with a further substitution $x_d \approx d$. In (1.1.3), we see the complex amplitude gain of the transmission.

The phase difference between the rays leads to partial cancellation, *in addition to* the d^{-2} free space loss. The received power is proportional to the expectation over the signal ensemble

$$E\left[\left(|\,\text{sum}(t)\,|\right)^2\right] = P_s \cdot \frac{1 - \cos(\beta \cdot \Delta x)}{d^2} \tag{1.1A.4}$$

where P_s is the power of the signal $s(t)$.

Let's see what this looks like. By inspection of the sketch above, we have the direct and reflected ray path lengths

$$x_d(d,h_b,h_m) := \sqrt{d^2 + (h_b - h_m)^2} \qquad x_r(d,h_b,h_m) := \sqrt{d^2 + (h_b + h_m)^2} \qquad (1.1A.5)$$

so that the differential path length is

$$\Delta x(d,h_b,h_m) := x_r(d,h_b,h_m) - x_d(d,h_b,h_m) \qquad (1.1A.6)$$

Set l to unity so that path difference is measured in wavelengths, and the average received power is

$$\lambda := 1 \qquad \beta := \frac{2 \cdot \pi}{\lambda}$$

$$P(d,h_b,h_m) := \frac{1}{d^2} \cdot \left[1 - \cos(\beta \cdot \Delta x(d,h_b,h_m))\right] \qquad (1.1A7)$$

Experiment with different antenna heights *in wavelengths*:

◧ Reference:D:\COURSES\MobChann\paperbook\Units.mcd(R)

$$h_b := 30 \qquad h_m := 5 \qquad\qquad \text{plot parameters:} \quad N := 100 \qquad k := 0 .. \frac{6}{4} \cdot N$$

$$d_k := 8 \cdot h_b \cdot h_m \cdot 1.05^{k-N}$$

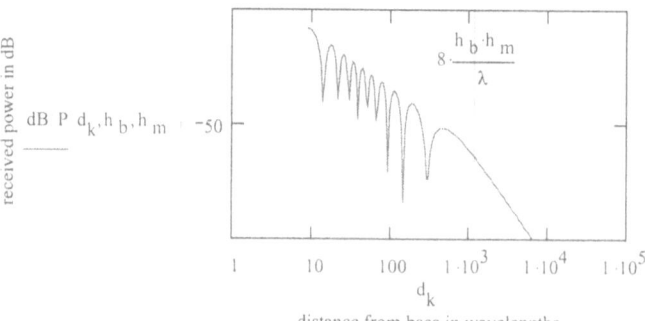

received power in dB

$\text{dB } P(d_k, h_b, h_m)$

distance from base in wavelengths

You can see that there are two regimes: near the base, there are strong ripples and the maxima drop according to inverse square law; and past the last maximum, the signal drops steadily as an inverse fourth power. You should also have found that there are twice as many maxima as there are wavelengths in the mobile antenna height. This is not surprising, since the differential path length decreases from $2h_m$, when the mobile is directly under the base antenna, to almost zero, when the mobile is far away from the base. If there are more than two rays, then there are more maxima

But why do we have two different regimes? For the answer, we resort to approximations and series expansions. Consider relatively distant behaviour, in which the separation distance d is much greater than the base and mobile antenna heights h_b and h_m. Using a first order expansion, we approximate the expressions given above for direct and reflected path lengths to obtain

$$x_d = \sqrt{d^2 + (h_b - h_m)^2} = d \cdot \sqrt{1 + \left(\frac{h_b - h_m}{d}\right)^2} \cong d \cdot \left[1 + \frac{1}{2} \cdot \left(\frac{h_b - h_m}{d}\right)^2\right] \quad \text{(approx)}$$

(1.1A.8)

$$x_r = \sqrt{d^2 + (h_b + h_m)^2} \cong d \cdot \left[1 + \frac{1}{2} \cdot \left(\frac{h_b + h_m}{d}\right)^2\right] \quad \text{(approx)}$$

(1.1A.9)

Therefore, the differential path length is

$$\Delta x = x_r - x_d \cong 2 \cdot \frac{h_b \cdot h_m}{d} \quad \text{(approx)}$$

(1.1A.10)

and the receive power is approximately

$$\frac{1}{d^2} \cdot P_s \cdot (1 - \cos(\beta \cdot \Delta x)) = \frac{1}{d^2} \cdot P_s \cdot \left(1 - \cos\left(\beta \cdot 2 \cdot \frac{h_b \cdot h_m}{d}\right)\right) \quad \text{(approx)}$$

(1.1A.11)

For small values of d, the cosine swings between -1 and +1, producing the close-in regime of $1/d^2$. However, in the distant regime, where the cosine argument is less than $\pi/2$ - that is, $d > 8h_b h_m/\lambda$ - we can use the series expansion for the cosine, giving the received power as

$$\frac{1}{d^2} \cdot P_s \cdot (1 - \cos(\beta \cdot \Delta x)) = \frac{1}{2} \frac{1}{d^2} \cdot P_s \cdot \left(\frac{4 \cdot \pi \cdot h_b \cdot h_m}{\lambda \cdot d}\right)^2 \quad \text{(approx)}$$

(1.1.12)

which varies as $\quad \dfrac{h_b^2 \cdot h_m^2}{d^4}$

Therefore the received power is inversely proportional to d^4, and is directly proportional to the squared antenna heights. The dependence on wavelength λ is more complicated than shown here, because atmospheric attenuation varies significantly with wavelength, as do the antenna power gains (for horn or planar antennas, the gains are inversely proportional to λ^2).

2. SHADOWING

In Section 1, we found that the exponent of the path loss between base and mobile varies between -3 and -4, giving rise to the inverse fourth power rule of thumb. In this section, we take a more local view, and consider the effect of major obstacles in the propagation path. As an example, the sketch below illustrates an unfortunate mobile with a hill between it and the base.

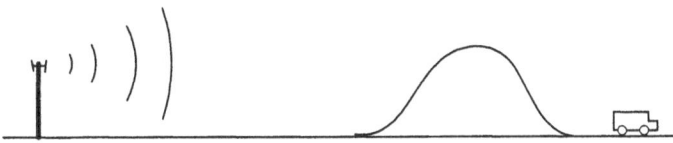

Typical obstacles are hills, large buildings, foliage, etc., and the resulting excess loss is termed *shadowing*.

Shadowing causes considerable variability about the mean power predicted by path loss. Because of the physical size of the obstacles that produce shadowing, the scale of significant variation is hundreds of wavelengths, and the shadow effect is roughly constant over many tens of wavelengths. However, if we average the signal strength around a circular path centred on the base, the familiar inverse cube or fourth power law reasserts itself.

2.1 Statistical Model of Shadowing

Shadowing causes the power to vary about the path loss value by a multiplicative factor that is usually considered as lognormally distributed over the ensemble of typical locales. An equivalent statement is that the power in dB contains a path loss term proportional to the log of distance, plus a Gaussian term with zero mean and standard deviation roughly 8 dB (for urban settings - rural topography exhibits a lower standard deviation).

2.2 Shadowing and System Design

Shadowing causes coverage holes, which may require fill-in by secondary transmitters. Shadowing also makes cell boundaries less well defined than simple path loss calculations suggest. In consequence, the algorithms controlling handoff, as a mobile progresses from one cell to the next, must be sophisticated, in order to avoid multiple handoffs and reversals of the handoff. Shadowing also shows up when we calculate cellular system capacity. We need an accurate model of the total cochannel interference from near and distant sources, and in this we must account for shadowing.

2.1 Statistical Model of Shadowing

Shadowing causes considerable variability about the mean power predicted by path loss. Because of the physical size of the obstacles that produce shadowing, the scale of significant variation is hundreds of wavelengths, and the shadow effect is roughly constant over many tens of wavelengths. However, if we average the signal strength around a circular path centred on the base, including many shadowing occurrences, the familiar inverse cube or fourth power law reasserts itself.

The graph below shows that shadowing produces fluctuations about an inverse third to fourth power path loss decrease. The fact that shadowing produces variations *above*, as well as below, the path loss may be confusing - after all, shadowing is an excess loss. The reason is simply that the path loss trend line is produced after the fact, as a best fit from measurements, so that the scatter is on both sides.

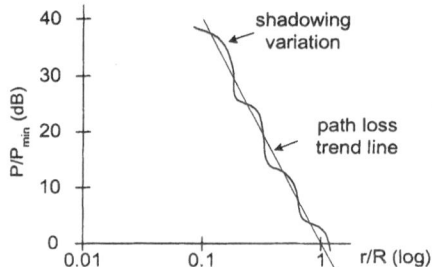

Propagation studies usually conclude that shadowing loss is well described by a lognormal distribution; when expressed in dB, the mean is 0 dB (because of the trend line noted above) and the standard deviation is typically 6 to 8 dB [**Stu96, Stee92, Yaco93**]. If you are unfamiliar with the log-normal distribution, see **Appendix A**.

There is a theoretical basis to the lognormal distribution in propagation studies: a typical signal has undergone several reflections or diffractions by the time it is received, each of which can be characterized by an attenuation, or multiplication. Expressed in dB, though, each event corresponds to subtracting a random loss. The result of these cascaded events is the sum of their dB losses which, by the central limit theorem, tends to converge to a normally distributed (Gaussian distributed) random variable. In natural units (i.e., not dB) this corresponds to a lognormal distribution.

If we combine the path loss and shadowing variations, the received power over a span of several tens of wavelengths at a distance r from the base is given by

$$P = P_0 \cdot \frac{10^{0.1 \cdot z}}{r^4} \qquad\qquad (2.1.1)$$

where z is normally distributed with mean 0 dB and standard deviation about 8 dB. Expressed in dB, the mean power over tens of wavelengths is

$$P_{dB} = P_{odB} - 40 \cdot \log(r) + z \qquad\qquad (2.1.2)$$

The shadowing level z depends on position, of course; if we measured the mean power over some other patch at the same distance r, we would find another value drawn from the same Gaussian (normal) ensemble, since different obstacles come into play.

Since z is a random variable, we are interested in how often it becomes so negative that the mean power P_{dB} becomes uncomfortably small and jeopardizes performance. Here we use the probability density function (pdf) of z. For unit standard deviation (i.e. 1 dB) it is the usual zero mean Gaussian (normal) density

$$p_z(z) := \frac{1}{\sqrt{2 \cdot \pi}} \cdot e^{-\frac{z^2}{2}}$$

We want its cumulative distribution function (cdf) - the probability that z is less than or equal to a given value

$$F_z(z) = \int_{-\infty}^{z} p_z(\zeta) d\zeta \qquad \text{or, using a Mathcad function,} \quad F_z(z) := \text{cnorm}(z)$$

Plot the pdf and cdf of the normal distribution for unit standard deviation

$$z := -3, -2.95 .. 3 \qquad t_5 := 1.645 \qquad t_{10} := 1.2815 \qquad t_{20} := 0.8417$$

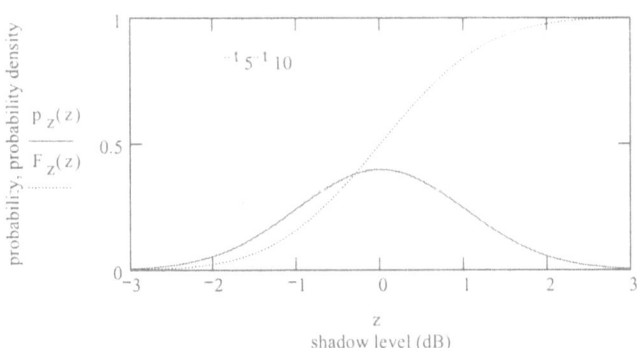

The values t_5, t_{10} and t_{20} are the points (measured in standard deviations) at which the cdf is 0.05, 0.10 and 0.20, respectively:

$$F_z \cdot t_5 = 0.05 \qquad F_z \cdot t_{10} = 0.1 \qquad F_z \cdot t_{20} = 0.2$$

Here's how to use these values. Suppose the shadowing standard deviation is 8 dB:

$$\sigma_{sh} := 8 \quad dB$$

Then the shadowing will be equal to or worse than

$$-t_5 \cdot \sigma_{sh} = -13.16 \quad \text{dB} \quad \text{in 5\% of the locations}$$

$$-t_{10} \cdot \sigma_{sh} = -10.25 \quad \text{dB} \quad \text{in 10\% of the locations}$$

$$-t_{20} \cdot \sigma_{sh} = -6.734 \quad \text{dB} \quad \text{in 20\% of the locations}$$

This suggests increasing the transmitter power by corresponding amounts to provide a safety margin against fading, with a given per cent reliability. We will do exactly this in the next section.

2.2 Shadowing and System Design

How does shadowing affect your life? If you are an RF planner, quite a bit. To lay out a new system, you work with a software package that accounts for the **propagation exponent** appropriate to the terrain and density of buildings, and for gross terrain features such as large hills. But that still leaves shadowing to disrupt this tidy picture. You'll see at least four problems caused by shadowing:

* noisy spots - poor coverage - along the cell edges;
* coverage holes within a cell;
* ragged handover between cells;
* poor carrier to interference ratio (CIR) in places;
* reduced system capacity caused by variable levels of interference from cochannel cells;

We'll have a brief look at them and some solutions below.

Shadowing and Edge of Cell Coverage

Suppose the combination of modulation, channel coding, source coding and grade of service dictates a minimum SNR of 20 dB. Clearly, you'll try to meet that value at the edge of the coverage area. But if you base your calculations only on path loss, you're in for a surprise. Shadowing will make the SNR less than the required value roughly half of the time! But how much less, and what to do about it?

How much less is a straightforward calculation, if we subscribe to the view that shadowing is lognormal and we know the standard deviation σ_{sh} of the shadowing We simply evaluate the statistics described in **Section 2.1**.

What to do about it is also straighforward - just allow a safety margin. Recall the values

$$t_5 := 1.645 \qquad t_{10} := 1.2815 \qquad t_{20} := 0.8417 \qquad\qquad (2.2.1)$$

from **Section 2.1**. Then if you want to be above the minimum SNR in 90% of the locations, and

$$\sigma_{sh} := 8 \quad dB$$

then add a margin of

$$\qquad\qquad (2.2.2)$$

$$t_{10} \cdot \sigma_{sh} = 10.252 \quad dB$$

$$\qquad\qquad (2.2.3)$$

That is, use your software to produce a SNR of $20 + t_{10} \cdot \sigma_{sh} = 30.252$ dB at the edge. Incidentally, use of this 10.3 dB value of shadowing margin is very common in system design.

Shadowing and Coverage Holes

Shadowing also produces coverage holes - areas in which a mobile finds its level may drop into the noise or interference. This is more a problem in the uplink, where transmit power from the mobile or cell phone is low, but we can also find holes near the edge of cells where the received power is low in both directions. Clearly, it is not desirable to have an area several hundred wavelengths wide with no service. One remedy for the larger holes (many of which are predictable by the RF propagation package and therefore don't really count as shadowing) is the use of an auxiliary antenna near or in the hole for fill-in.

In the uplink direction, where the mobile transmits and the auxiliary antenna is used for reception, this approach works well. The signal from the auxiliary antenna is fed back to the base by landline or microwave, where the receiver can make use of both the auxiliary and its own signals. It either selects the stronger one, or, with more difficulty, combines the two before detection. In the downlink direction, where the auxiliary antenna transmits to mobiles in the coverage hole, there are practical complications. We not only have to find a site for the antenna, as we do for reception, but we have to ensure that the transmission of RF is safe for those living nearby (fortunately, the transmit power can be low).

In addition, use of auxiliary antennas in the downlink we can do more harm than good to the signals if we are not careful. For example, if the carrier frequencies of the main and auxiliary signals are slightly offset, then the total signal received by the mobile beats at the difference frequency. Also, differences in propagation distance and delay may become comparable to the time scale of the modulation, resulting in delay spread and signal distortion.

Despite the difficulties associated with transmission from auxiliary antennas, it is frequently used, and is termed *simulcast* or *quasi-synchronous transmission*. In fact, the separate signals can be a valuable form of diversity as protection against fading, as long as they can be distinguished. (This will become clearer in the following sections). For example, the POCSAG paging system [**Brit78**] transmits duplicate signals from the base and auxiliary antennas. Consequently, a pager that might be in a deep fade if there were only one transmitting antenna may receive a good signal from the other antenna. The antenna signals deliberately have a slight carrier offset, to produce beating. For this reason, the two signals cannot destructively interfere for more than half the beat period, which is set to half the symbol duration. In this way, we exchange possibly long-lasting fades in space with brief fades in time. In CDMA, an alternative is to keep the same carrier, but delay the spreading code in one of the transmissions [**Sous95**]. In effect, we create an artificial, and separable, diversity structure.

Shadowing and Handoff

Because shadowing causes the edges of cells to become only approximately defined, it produces difficulty in the handoff of a mobile from one base station to another in the next cell. If a coverage hole is near the edge of a cell, the signal from the neighbouring cell may be stronger - but not for long. If a handoff is initiated prematurely, it may have to be reversed, then reversed again. The result is a flurry of extra protocol exchanges over the air and over the underlying network, causing unnecessary system load. The sketch below shows what can happen as the mobile crosses a cell boundary. The "islands" in each cell where the other cell's transmitter is stronger could result in five handoff actions, instead of one, if we do not address the problem.

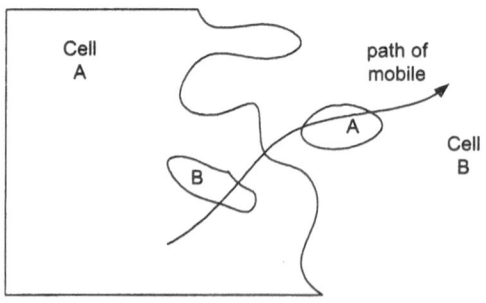

There have been many studies of handoff in the literature; e.g., [**Vija93**]. Many of them deal with selection of the threshold power difference between signals the mobile receives from different cells and the use of some form of hysteresis to prevent repetitive handoffs back and forth.

Shadowing and Cochannel Interference

Back in **Section 1.2**, we calculated C/I in a cellular system using only the path loss. Here's the sketch again, as a reminder.

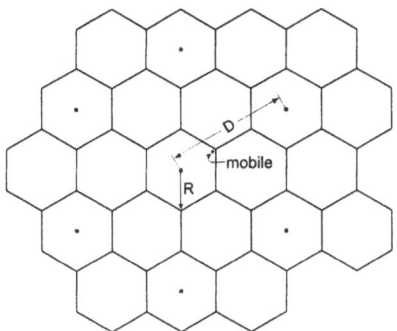

All the signals received by the mobile are affected by shadowing. Therefore the total cochannel interference (CCI) power from the first cochannel tier (i.e. the closest 6 cochannel cells), given by

$$I = \sum_{k=1}^{6} I_k \qquad (2.2.4)$$

is randomly distributed about a mean value determined by the simple inverse fourth power path loss calculations. If we want to know the probability that the C/I goes below some threshold of acceptability (and we do!) then we need the statistics of both C and I. Here we have good news and bad news. The bad news is that there is no closed form for the pdf of the sum of lognormally distributed random variables I_k. The good news, though, is that it can be approximated reasonably as a lognormally distributed random variable itself, if the individual pdfs are identical.

If the sum of lognormal variates (the total CCI) is to be approximated by a lognormal pdf, all we need to select are an appropriate mean and variance, the two parameters of the pdf. Unfortunately, how to do it is a vexed question. For example, we might note that, since the I_k in (2.2.4) are independent, their sum I has mean equal to the sum of the means, and variance equal to the sum of the variances. An approximation to the pdf of I might therefore be selected to have the same first and second moments. This is known as Wilkinson's method. Unfortunately, it is a poor approximation on the tails of the pdf if σ is large and there are many terms in the sum. Several other selection methods have been proposed, all having merits in different regions of σ, number of terms and how far out on the tail of the pdf we need to go. For a good comparison, see [**Beau95**] or [**Stub96**]. For a good analysis of the effect on system behaviour, see [**Yeh84**].

In the special case of a single interfering signal, the distribution of C/I and the safety margin are obtained more easily. The C/I is

$$dB\left(\frac{C}{I}\right) = dB\left(\frac{C_o}{I_o}\right) + z_C - z_I \qquad\qquad (2.2.5)$$

where the subscript o denotes nominal values determined by path loss calculations. The two shadowing values are normally distributed, since they are in dB, so their difference is also normal, and also has zero mean. They are independent, since the signal and interference usually arrive from different directions (they are usually from transmitters in different cells), and the variance of the difference is therefore the sum of the variances. Replace the difference $z_C - z_I$ with a single shadowing value z_R for the ratio, and note that it has mean and variance

$$\mu_R := 0 \qquad \sigma_r := \sqrt{2} \cdot \sigma_{sh}$$

For a reliability of 90%, then, we need to allow a margin of

$$t_{10} \cdot \sigma_r = 14.499 \quad \text{dB, assuming individual shadowing with standard deviation } \sigma_{sh} = 8$$
$$\text{dB.}$$

3. FADING AND DELAY SPREAD

Our path loss models of **Section 1** considered only large scale variations of signal power, where an entire cell, possibly several kilometres in radius, was the domain of interest, and all points with the same radius were considered to have the same path loss. The phenomenon of shadowing in **Section 2** took a more local view of power variations, where the variations of interest were hundreds of wavelengths across. We considered shadowing as a random variation of power about a mean that determined by the path loss model.

In this section, we take an even more local view and examine fading, the interference among many scattered signals arriving at an antenna. Here the scale is very small, with level changes of tens of dB and violent phase swings across a fraction of a wavelength. We consider this fading to be a random fluctuation about the mean power level determined by the path loss and shadowing models. We expect the fading statistics to be reasonably stationary over distances of a few tens of wavelengths.

Fading has been studied intensively for several decades, because it is so destructive to reliable communication. Many methods have been developed to mitigate its effects - and we shall look at several of them in the other texts in Mobile and Personal Communications in this series - but it remains one of the two major problems in the area (the other being interfering transmitters). That's why most of the rest of this course is taken up with various aspects of fading.

Fading phenomena are not confined to mobile communications - they occur any time a narrowband signal is received from a scattering medium. Thus the modeling and analysis in this section also applies directly to:

* laser speckle noise
* sonar returns
* scintillation in radar returns from icebergs and floes
* acoustics of ultrasound

Below are summaries of the subsection contents, with links to the detailed discussions.

3.1 Physical Basis of Fading

The receiver antenna picks up multiple reflections of the transmitted signal, each with its own amplitude, phase and delay. They can interfere constructively or destructively, depending on the antenna position, producing a random standing wave. There can be major changes in amplitude, phase and filtering of the resultant over distances as short as half a wavelength.

3.2 Mathematical Model of Fading

Normally, we transmit relatively narrowband signals, in which the carrier and the modulation are considered separately. In this section, we'll obtain a straightforward model of the effects of the multipath channel on the complex envelope (i.e., the modulation) of the signals.

3.3 Consequences: Doppler Spread, Delay Spread and Time Variant Filtering

As the mobile moves through the standing wave, the amplitude, phase and filtering applied to the transmitted signal change with time at a rate proportional to vehicle speed. If the transmitted signal is unmodulated carrier, the output is time-varying and therefore has a nonzero spectral width. This is Doppler spread. If the transmitted signal is an impulse, the delays of the reflections cause it to be received spread out in time. This is delay spread. Doppler spread and delay spread are generated by two different means, and either or both can be present or absent in typical mobile situations. One way to describe the combination of the two is a time-variant linear filter.

3.4 Is it Flat or Frequency Selective? An Example

This example shows how the product of delay spread and signal bandwidth determines whether a channel is flat or frequency selective. The severity of its effect depends on the modulation, of course, but these graphs should give you an intuitive understanding of what you're up against.

3.1 Physical Basis of Fading

The sketch below shows a typical link between mobile and base station antennas. There are several reflectors - buildings, hills, other vehicles, etc. - around the mobile, but few or none near the base station, because it is usually mounted high above its surroundings. These reflectors are known generically as scatterers. Communication between the base and mobile takes place over many paths, each of which experiences one or more reflections, and the receiver picks up the sum of all the path signals. Note that this applies to either direction of transmission, to or from the mobile.

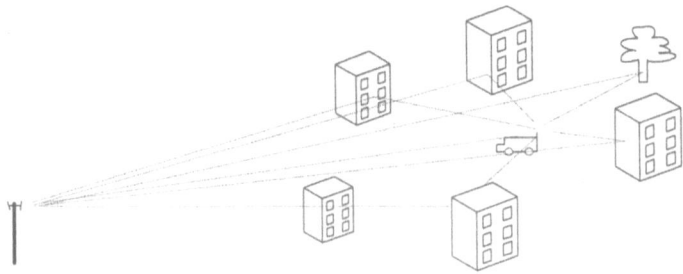

We can infer much of what happens just from consideration of the sketch.

* Since the individual paths are linear (i.e., they satisfy the superposition requirements), the overall multipath channel is linear.

* Each path has its own delay and gain/phase shift, so the aggregate of paths can be described by its impulse response or frequency response. Therefore different carrier frequencies will experience different gains and phase shifts. ("Gain" is used in a general sense here, since the paths really experience attenuation.)

* Whether the range of delays (the "delay spread") has a significant effect on the modulation of the carrier depends on the time scale of the modulation (roughly, the reciprocal of its bandwidth). This implies that the dimensionless product of channel delay spread and signal bandwidth is an important measure.

* If the mobile changes position, the paths all change length in varying amounts Since a change in path length of just one wavelength produces 2π radians of phase shift, a displacement of a fraction of a wavelength in any direction causes a large change in the aggregate gain and phase shift, as the sum of the paths shifts between reinforcement and cancellation.

* When the mobile moves through this two-dimensional standing wave pattern, the impulse response and frequency response change with time, so the channel is a time-varying linear filter. The time variant nature of the net gain is termed "fading" and the fastest rate of change is the "Doppler frequency".

* Whether the time-varying nature of fading has a significant effect on the modulation of the carrier depends on the time span of the required receiver processing (e.g., differential detection over two symbols, equalization over many symbols, etc.). The dimensionless product of this time span and the Doppler frequency is another important parameter.

The standing wave pattern around the mobile is not unlike ripples in water, where reflections from nearby rocks and logs interfere to create an area of complex disturbance, where successive maxima are spaced very roughly by half a wavelength, without really being periodic. In later sections, we will use random process models to describe the pattern, but for now you can learn more simply by looking at it. The **random field example** in Appendix C helps you visualize the standing wave (its magnitude, at least).

A more common picture is the signal strength in dB along an arbitrary direction for an unmodulated carrier. Here we use the **Jakes' complex gain generator** for isotropic scattering in Appendix B:

⊡ Reference:D:\COURSES\MobChann\paperbook\Jakesgen.mcd(R)

$$dB(x) := if\left(x < 10^{-14}, -140, 10 \cdot \log(x)\right)$$

select M points spaced by 0.01λ

$\lambda := 1$ $M := 400$ $i := 0 .. M$ $x_i := 0.01 \cdot \lambda \cdot i$

To see other fading signals, put the cursor on the initialization below and press F9:

$G := \text{Jakes\_init}(0)$ $g_i := \text{Jakes\_gen}(x_i, G)$

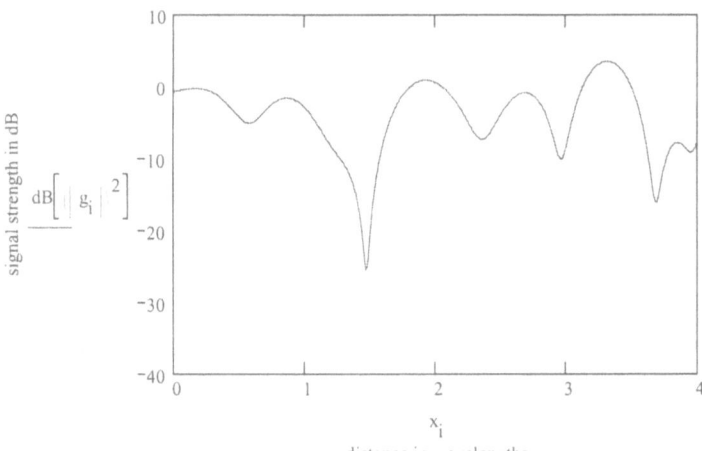

Signal Strength as Function of Position

The graph shows tens of dB variation over a fraction of a wavelength. It also shows a rough periodicity in the received signal strength, with a roughly $\lambda/2$ spacing.

We have identified two multipath phenomena: fading and delay spread. Either or both can be significant or not in a given terrain with a given modulation rate. For example, we expect fading to be rapid with vehicular use, but modest with pedestrian use. Similarly, delay spread can be large in hilly or mountainous terrain, but smaller in an urban core, where multiple reflections quickly attenuate the signals with longer path lengths. The severity of the two impairments depends on the modulation. Consider, for example, a typically short urban delay spread of 3 μs and the fast fade rate of 200 Hz. Differentially encoded BPSK at 2400 b/s would not be affected much by the delay spread, but would suffer from pulse distortion because of the rapid fading. Conversely, the same modulation format at, say, 200 kb/s would certainly require equalization because of the delay spread, but the short pulses would experience no significant distortion from the time variations of the fading.

Finally, a link back to our earlier propagation models of path loss and shadowing. Recall that they are typically modeled as inverse fourth power and randomly lognormal, respectively. In fading, we have variation on a still smaller spatial scale. Thus we have a set of spatially nested models:

* **Path loss** is due to a combination of inverse square law and destructive interference at grazing angles. The simplified model asserts that path loss depends only on distance from the base station, so that all points on a circle centred on the base have the same received power.

* **Shadowing**, produced by obstructions such as buildings, trees or small hills, introduces more realism. Now points on a circle centred on the base have received power that may be greater or less than the value predicted by path loss calculation, depending on the local environment. The scale of significant variation is hundreds of wavelengths and we expect the received power to be more or less constant over a patch tens of wavelengths across. In Section 2, **equation (2.1.1)** showed that the received signal power in such a patch is proportional to the product of path loss and shadowing attenuations.

* Fading - the microstructure of signal level variation - is due to interference among the multiple paths, with significant variations over about $\lambda/2$. When we average the power over the many local fades in a patch tens of wavelengths across, we identify the result as the average power predicted by path loss and shadowing.

3.2 Mathematical Model of Fading

Radio signals are always bandpass, and are almost always narrowband. In this case, we can describe the channel effects in terms of the modulation, as distinct from the carrier. This section has two objectives:

* to develop a mathematical model for the effects of the multipath channel on the complex envelope of the signal;

* to extend the model to include the effects of mobile motion.

Effects on the Complex Envelope

The transmitted bandpass signal at carrier frequency f_c with complex envelope $s(t)$ is denoted by

$$s'(t) = \text{Re}\left(s(t) \cdot e^{j \cdot 2 \cdot \pi \cdot f_c \cdot t}\right) \tag{3.2.1}$$

It is transmitted in a multipath environment, as shown by the sketch of the mobile surrounded by scatterers and moving at velocity v.

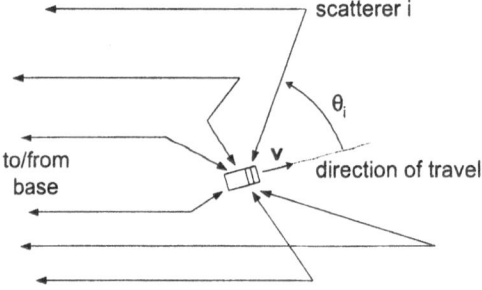

The individual paths, indexed by i, have path length x_i and reflection coefficient a_i, which may represent the combined effect of several reflections. The received bandpass signal is therefore the superposition

$$y'(t) = \sum_i a_i \cdot s'\left(t - \frac{x_i}{c}\right) = \sum_i a_i \cdot \text{Re}\left[s\left(t - \frac{x_i}{c}\right) \cdot \exp\left[j \cdot 2 \cdot \pi \cdot f_c \cdot \left(t - \frac{x_i}{c}\right)\right]\right]$$

$$\tag{3.2.2}$$

$$= \text{Re}\left[\sum_i a_i \cdot s\left(t - \frac{x_i}{c}\right) \cdot \exp\left[j \cdot 2 \cdot \pi \cdot \left(f_c \cdot t - \frac{x_i}{\lambda}\right)\right]\right]$$

where c is the speed of light, and the wavelength is $\lambda = c/f_c$. We can now factor out the carrier $\exp(j2\pi f_c t)$ and write the received signal in complex envelope terms

$$y'(t) = \text{Re}\left(y(t) \cdot e^{j \cdot 2 \cdot \pi \cdot f_c \cdot t}\right) \tag{3.2.3}$$

where the received complex envelope is the sum of variously attenuated, phase shifted and delayed path components

$$y(t) = \sum_i a_i \cdot e^{-j \cdot 2 \cdot \pi \cdot \frac{x_i}{\lambda}} \cdot s\left(t - \frac{x_i}{c}\right) = \sum_i a_i \cdot e^{-j \cdot 2 \cdot \pi \cdot f_c \cdot \tau_i} \cdot s\left(t - \tau_i\right) \tag{3.2.4}$$

and the time delays are $\tau_i = x_i/c$. In (3.2.4) we have the desired complex envelope model.

In some cases, there is a line of sight (LOS) path to the mobile, as well as the scattered paths. The LOS path will be the first to arrive, since the others have more indirect paths to traverse. It is usually the strongest of the individual paths, but it is not necessarily stronger than the aggregate of the scattered paths.

The Effect of Motion

Next, we'll consider the effect of motion in the basic description (3.2.4). When the mobile moves through this welter of arriving reflections, the path lengths change. If the angle of arrival of path i with respect to the direction of motion is θ_i, then the path length change, as a function of speed v and time t, is

$$\Delta x_i = - v \cdot \cos\left(\theta_i\right) \cdot t \tag{3.2.5}$$

This shifts the frequency of each component by an amount dependent on its arrival angle θ_i. Here's why. The complex envelope of the channel output is, from (3.2.4),

$$y(t) = \sum_i a_i \cdot e^{-j \cdot 2 \cdot \pi \cdot \frac{(x_i + \Delta x_i)}{\lambda}} \cdot s\left[t - \frac{(x_i + \Delta x_i)}{c}\right]$$

$$\tag{3.2.6}$$

$$= \sum_i a_i \cdot e^{-j \cdot 2 \cdot \pi \cdot \frac{x_i}{\lambda}} \cdot e^{j \cdot 2 \cdot \pi \cdot \frac{v}{\lambda} \cdot \cos \theta_i \cdot t} \cdot s\left(t - \frac{x_i}{c} + \frac{v \cdot \cos \theta_i \cdot t}{c}\right)$$

We can simplify this. First, include the phases $2\pi x_i/\lambda$ in the phase of a_i. Next, since the changes in signal delay $v\cos(\theta_i)t/c$ are small compared with the time scale of the modulation $s(t)$, ignore them. The result is

$$y(t) = \sum_i a_i \cdot e^{j \cdot 2 \cdot \pi \cdot \frac{v}{\lambda} \cdot \cos(\theta_i) \cdot t} \cdot s\left(t - \frac{x_i}{c}\right) = \sum_i a_i \cdot e^{j \cdot 2 \cdot \pi \cdot f_D \cos(\theta_i) \cdot t} \cdot s(t - \tau_i) \qquad (3.2.7)$$

showing that the ith scatterer shifts the input signal in time by τ_i and in frequency by $f_D \cos(\theta_i)$. The maximum Doppler shift is

$$f_D = \frac{f_c}{c} \cdot v = \frac{v}{\lambda} \qquad (3.2.8)$$

Equation (3.2.7) and its static counterpart (3.2.4) are fundamental descriptions of a channel with discrete scatterers.

Finally, we'll modify the discrete scatterer model (3.2.7) for the case of a large number of scatterers. In this case, it is often more convenient to replace the sum with an integral over a density. Define the *delay-Doppler spread function* by

$$\gamma(v, \tau) \cdot dv \cdot d\tau = \sum_{i'} a_{i'} \qquad (3.2.9)$$

where i' indexes all scatterers with delay in $d\tau$ at τ and Doppler in dv at v. (Note that frequency v (nu) in Hz is a different symbol from v (speed)). Then the continuous equivalent of (3.2.7) is

$$y(t) = \int_0^\infty \int_{-f_D}^{f_D} \gamma(v, \tau) \cdot e^{j \cdot 2 \cdot \pi \cdot v \cdot t} \cdot s(t - \tau) \, dv \, d\tau \qquad (3.2.10)$$

We'll see the delay-Doppler spread function again in the discussion of linear time varying filters in **Section 3.4** and the discussion of WSSUS channels in **Section 5.3**.

Incidentally, if you want to return to the discrete scatterer model from (3.2.10), just define

$$\gamma(v, \tau) = \sum_i a_i \cdot \delta(v - f_D \cdot \cos(\theta_i)) \cdot \delta(\tau - \tau_i) \qquad (3.2.11)$$

Substitution into (3.2.10) produces (3.2.7) again.

3.3 Consequences: Doppler Spread, Delay Spread and Time Variant Filtering

Section 3.1 introduced a physical model of the mobile channel based on multiple reflections. Section 3.2 translated it into a simple mathematical model of multiple arrivals, each with its own amplitude, phase, delay and Doppler shift.. It was summarized in the input/output relation (3.2.7), reproduced here as

$$y(t) = \sum_i a_i \cdot e^{j \cdot 2 \cdot \pi \cdot f_D \cdot \cos(\theta_i) \cdot t} \cdot s(t - \tau_i) \qquad (3.3.1)$$

This section explores some of the consequences. They are:

  * fading (or time selective fading, to be precise), which is linked to Doppler spread produced by vehicle motion,

  * and delay spread, which is linked to frequency selective fading.

They are produced by two separate mechanisms, and either of them can be present or absent in common mobile situations. This section also introduces the time-variant filter view of the mobile channel, although this theory is treated carefully only in the optional Section 3.4.

By the way, we often ignore the mean delay in (3.3.1) and focus instead on differential delays. However, the mean path length also changes as the vehicle moves, illuminating new scatterers and leaving others behind.

Doppler Spread

The first phenomenon is fading, or Doppler spread. If the mobile moves through the random field, then it experiences changes in signal level and phase, with the rate of changes proportional to the mobile speed. It's easiest to start with *flat fading*, in which the signal bandwidth is so small that the delays τ_i in (3.3.1) do not affect the signal, and $s(t-\tau_i) \approx s(t)$. Then from (3.3.1),

$$y(t) = s(t) \cdot \sum_i a_i \cdot e^{j \cdot 2 \cdot \pi \cdot f_D \cdot \cos(\theta_i) \cdot t} = g(t) \cdot s(t) \qquad (3.3.2)$$

where the *channel complex gain* $g(t)$ is time varying because the phase angles $2\pi f_D cos(\theta_i)t$ change with time. In other words, the Doppler shifts and the variation of gain with observation time are produced by the same means (motion of the mobile) and can be considered alternative descriptions of the same phenomenon.

This time-varying complex gain is brought to life in the **animation of complex gain** of Appendix D - don't miss it! You'll see that $g(t)$ becomes close to zero from time to time, corresponding to a fade of the signal $y(t)$. Since $y(t)$ varies with time rather more quickly than does $s(t)$, the received signal is somewhat spread out in frequency from the transmitted signal - hence the term Doppler spread. In particular, a pure tone, $s(t)=A$, will spread out to several components in the band $[-f_D, f_D]$ (one for each scatterer); that is, an impulse in frequency spreads to a finite width.

As rules of thumb, typical values of maximum Doppler f_D at 1 GHz are about 100 Hz for vehicle use, 3 Hz for pedestrian use and (of course) 0 Hz when the unit is motionless in a parked car or sitting on a desk. The latter value is troublesome, because the unit could be in a deep fade for prolonged periods. Doubling the carrier frequency to 2 GHz doubles the Doppler spread and fade rate, since $f_D=v/\lambda$, where v is the vehicle speed.

How much the multiplicative distortion $g(t)$ in (3.3.2) affects the received signal depends on the fade rate and the time span of the required receiver processing. That time span may be one symbol for coherent detection, two symbols for differential detection or a few symbols for an equalizer. If the complex gain does not change significantly over this time, then the primary effect is just a slowly varying SNR. We term this condition *slow fading*, and the criterion is that the product of fade rate and processing window is very small: $Nf_DT \ll 1$, where N is the receiver processing window measured in symbols and T is the symbol duration. On the other hand, if this product is significant, so that the signal in the window, or even individual data pulses, are distorted, it is termed *fast fading*. This is a big problem: the distortion, especially the phase hits you saw in the **animation of complex gain**, produces errors, even if there is no noise. This the *irreducible error rate* or *error floor*. No increase in transmit power will eliminate it.

Delay Spread

The second phenomenon is delay spread. For very small Doppler or a stationary mobile, we can consider the phases of the reflections to be constant. Adapting (3.3.1) above, we have

$$y(t) = \sum_i g_i \cdot s(t - \tau_i) \qquad \text{where} \qquad g_i = a_i \cdot e^{-j \cdot \phi_i} \qquad (3.3.3)$$

where ϕ_i is a random phase associated with the arrival. It is clear that the channel is a linear finite impulse response (FIR) filter with impulse response

$$\sum_i g_i \cdot \delta(t - \tau_i) \qquad (3.3.4)$$

If the transmitted signal is an impulse in time, the reflections spread it out upon reception - hence the term delay spread. The sketch below of an impulse response seen at some time t_o shows the g_i as real, for simplicity.

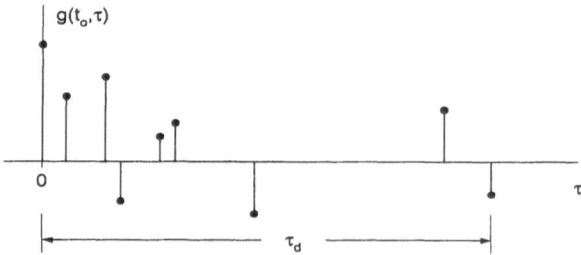

The corresponding frequency response and its effect on the complex envelope are

$$G(f) = \sum_i g_i \cdot e^{-j \cdot 2 \cdot \pi \cdot f \cdot \tau_i} \qquad\qquad Y(f) = G(f) \cdot S(f) \qquad\qquad (3.3.5)$$

The range of delays τ_d, called the *delay spread*, causes variation of $G(f)$ across the signal band. In other words, the delay shifts and the variation of gain with input frequency are produced by the same means (different path lengths among the scatterers) and can be considered alternative descriptions of the same phenomenon.

Significant variation of $G(f)$ across the band produces signal distortion. For digital transmission, this shows up as intersymbol interference (ISI), which may require equalization. FIR filters, like our channel, may have deep nulls at certain frequencies, so that part of the signal can be notched out. This is termed *frequency selective transmission*. In contrast, if the range of delays is not significant, then the channel is considered *flat*, since there is little variation of frequency response across the signal band. Neither condition implies changes over time, since the mobile could be stationary.

What criterion distinguishes flat transmission from frequency selective transmission? Clearly it depends on the signal, as well as the channel. If the delay spread is very small compared with the fine structure in time of the signal (i.e., compared with the reciprocal bandwidth of the signal), then it is flat; that is, $\tau_d W \ll 1$, where W is the bandwidth of the signal.

Frequency Selective Fading

You have seen two distinct phenomena in this discussion: fading and delay spread. They arise from different mechanisms, and they can appear separately or together. To understand mobile communications, you must distinguish between them clearly.

What if the two phenomena are both present? If the mobile moves through a random field in which the delay spread cannot be ignored, given the signal bandwidth, then from (3.3.1)

$$y(t) = \sum_i g_i(t) \cdot s(t - \tau_i) \qquad \text{where} \qquad g_i(t) = a_i \cdot e^{j \cdot 2 \cdot \pi \cdot f_D \cdot \cos(\theta_i) \cdot t} \qquad (3.3.6)$$

We now have both Doppler spread and delay spread. This is termed *frequency selective fading*, and it gives rise to real challenges in modulation and detection, especially if the fading is fast.

This channel is a time variant linear filter. As background, recall the interpretation of impulse response (often denoted $h(\tau)$) in a linear time invariant system: first, it is the response observed at some arbitrary time t to an impulse τ seconds earlier, at t-τ; and second, it does not depend on the observation time t - only on the delay τ between the times of application of the impulse and observation of the result. In contrast, (3.3.6) shows that our channel has an impulse response that depends on observation time t, as well as the delay τ:

$$g(t, \tau) = \sum_i g_i(t) \cdot \delta(\tau - \tau_i) \qquad\qquad (3.3.7)$$

where $g(t,\tau)$ is the response observed at time t to an impulse applied τ seconds earlier; that is, at time t-τ. The familiar linear time invariant filter is obtained if $g(t,\tau)$ does not depend on observation time t, and becomes just $g(0,\tau)$.

The input-output relationship in the time domain is slightly more complicated than the usual convolution, and is given by

$$y(t) = \int_0^\tau d\; g(t,\tau) \cdot s(t-\tau) \, d\tau \qquad (3.3.8)$$

The frequency-domain counterpart of (3.3.7) is

$$y(t) = \int_{-\infty}^\infty G(t,f) \cdot S(f) \cdot e^{j\; 2 \cdot \pi \cdot f \cdot t} \, df \qquad (3.3.9)$$

where $G(t,f)$ is the time-variant frequency response; that is, the gain at observation time t experienced by a complex exponential $\exp(j2\pi ft)$ at frequency f. You have already seen a third input-output relationship in **Section 3.2**:

$$y(t) = \int_0^\infty \int_{-f_D}^{f_D} \gamma(v,\tau) \cdot e^{j\; 2 \cdot \pi \cdot v \cdot t} \cdot s(t-\tau) \, dv \, d\tau \qquad (3.3.10)$$

where $\gamma(v,\tau)$ is the delay-Doppler spread function and v is the frequency shift $f_D \cos(\theta)$.

Summary

The basic model for fading in mobile communications is linear, and involves both time variation and filtering. The two phenomena at work are:

     * Fading (or time selective fading) refers to variation in the channel response with observation time, and it is linked to the Doppler shifts the signal experiences on the various paths. It is caused by motion of the mobile's antenna through the complicated standing wave produced by reflections from scatterers.

     * Frequency selective transmission refers to variation in the channel response with input frequency, and it is linked to the delays the signal experiences on the various paths. It is caused by the different path lengths imposed by locations of the scatterers.

When both phenomena are present, we have frequency selective fading, and input-output relations that are described by time-variant linear filters.

Appendix I provides more information on time-variant filters, if you're curious. The two traditional input-output relations - convolution in time and multiplication in frequency - are extended to two dimensions and four I/O relations. Those relations are: time variant impulse response; time variant transfer function; output Doppler spread function; and delay-Doppler spread function. You should become comfortable with the first two - time variant impulse response and time variant transfer function - because they arise frequently in technical papers and analyses, and even in your simulations. The third one is interesting for its simplicity, but it is little used. As for the delay- Doppler spread function, it is a useful conceptual tool - it arose naturally from the physical model in **Section 3.1** and its mean square value makes an important and fundamental appearance in **Section 5.3** as the scattering function.

Model Criteria

Finally, we summarize the models and criteria that relate them in the diagram below. Note that the classification of channels depends on the time scale of the signal. To be specific, assume that it is a data signal with symbol period T and the receiver uses coherent detection, with a one-symbol processing window.

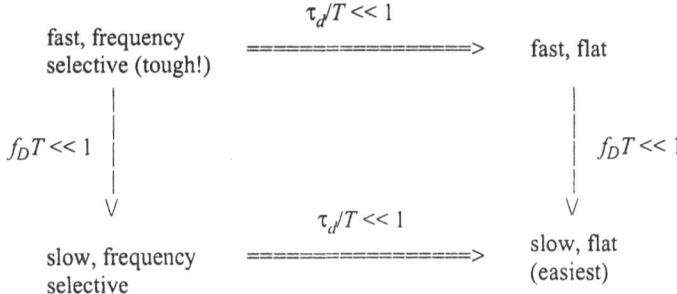

This diagram indicates the consequences of your choice of signaling rate. The faster you transmit (smaller T), the greater the signal bandwidth becomes, and the more frequency selective the channel appears over that bandwidth. At some point, you have to deal with intersymbol interference using an equalizer or equivalent. On the other hand, the more slowly you transmit, the faster the channel becomes (in relative terms), and you have to deal with pulse distortion and phase decorrelation from

pulse to pulse. Clearly, you would like $f_D\tau_d \ll 1$, in order to give you some room to select an appropriate signaling rate. If that doesn't hold, you are stuck with fast, frequency selective fading and a difficult design.

A Treat

Now for a treat - you've earned it by getting this far! Have a look at the **animated frequency response** in Appendix E. It shows how the magnitude and group delay of the frequency response change as the mobile moves through the random standing wave, causing the filter coefficients to change.

3.4 Is it Flat or Frequency Selective? An Example.

As noted in **Section 3.3**, whether the channel is flat or frequency selective depends on the bandwidth of the signal - that is, how quickly it varies in time - and on the delay spread. A numerical example is instructive here. You can choose the delay spread in a two-ray channel and see whether the variation in frequency response over the band is significant. The phases of the two rays are randomized.

The example signal $s(t)$ is a single sinc pulse, so it has a rectangular spectrum with lowpass bandwidth $W/2$:

$$W := 1 \qquad \text{sinc}(x) := \text{if } x{=}0, 1, \frac{\sin(\pi \cdot x)}{\pi \cdot x} \tag{3.4.1}$$

$$s(t) := W \cdot \text{sinc}(W \cdot t) \qquad S(f) := \left| f \right| \le \frac{W}{2} \tag{3.4.2}$$

To keep the demonstration simple, consider a channel of only two rays, with random phases and random amplitudes in $[0,1]$:

$$i := 0..1 \qquad g_i := e^{j \cdot \text{rnd}(2 \cdot \pi)} \cdot \text{rnd}(1) \tag{3.4.3}$$

If you want to see the effect of other choices of the phases and amplitudes, put the cursor on the g_i equation and press F9. You'll get newly randomized phases. Next, you choose the delay spread, giving the path delays and frequency response as shown:

$$\tau_d := .05 \qquad\qquad \tau_i := \tau_d \cdot i \qquad G(f) := \sum_i g_i \cdot e^{-j \cdot 2 \cdot \pi \cdot f \cdot \tau_i} \tag{3.4.4}$$

so the fractional delay spread: $\tau_d \cdot W = 0.05$ <=== this is the
                                                              criterion

The group delay is readily calculated by means of the logarithmic derivative:

$$\tau_{gp}(f) := \frac{-1}{2 \cdot \pi} \cdot \text{Im}\left[\frac{\sum_i -j \cdot 2 \cdot \pi \cdot \tau_i \cdot g_i \cdot e^{-j \cdot 2 \cdot \pi \cdot f \cdot \tau_i}}{G(f)} \right] \tag{3.4.5}$$

The received signal is

$$y(t) := \sum_i g_i \cdot s(t - \tau_i) \qquad Y(f) := S(f) \cdot G(f)$$

The frequency response is plotted in magnitude and in group delay below. See how the graphs change when you:

* randomize the phase angles again in (3.4.3) with F9 (or from the menu Math/Calculate Worksheet);
* change the delay spread in (3.4.4).

plot range: $f := -2, -1.95 .. 2$ $\tau_d \cdot W = 0.05$

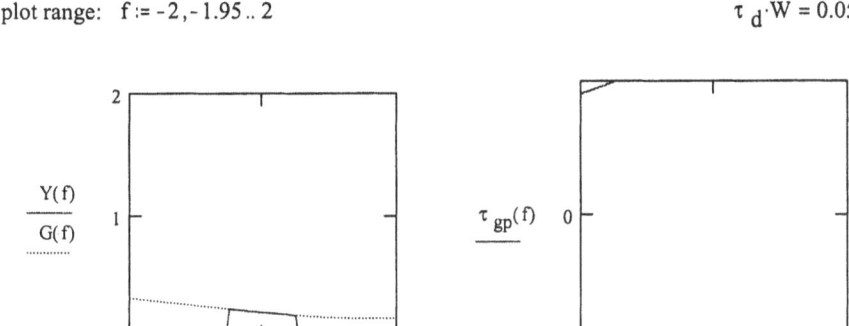

$\dfrac{Y(f)}{G(f)}$ $\tau_{gp}(f)$

Mags of Freq Resp and Output   Group Delay

You can see that, for small values of $\tau_d W$, the frequency response is almost flat across the signal band, with little group delay variation. That means very little signal distortion, even though if the path phases are almost opposite and the gains are roughly equal, then we have a deep fade across the band. On the other hand, for larger values of $\tau_d W$, we see frequency selective transmission and, consequently, significant signal distortion. Usually, not all the signal band is deeply faded - some part of the signal energy gets through. This fact is exploited in CDMA and, in fact, in the humble equalizer. It's a hard-won form of diversity.

Next, we look at the response in the time domain. Again, you can change the phases or delay spread and observe the changes in selectivity.

$$t := \frac{2}{W}, \frac{1.9}{W} .. \frac{3}{W} \qquad \text{time range for plot}$$

First, the impulse response (red for real component, blue for imaginary):

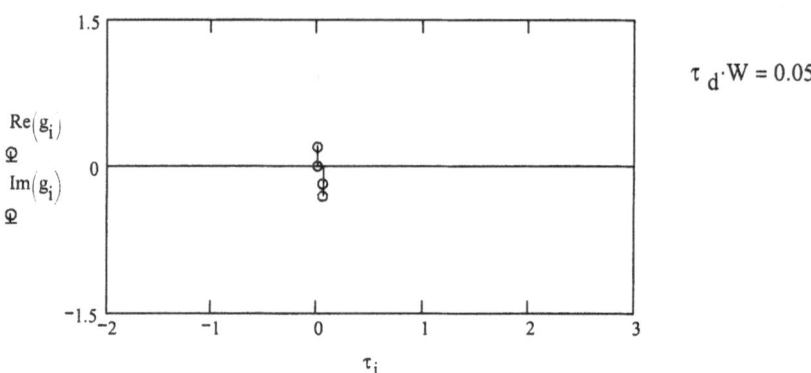

$\tau_d \cdot W = 0.05$

Again, see how the graphs change when you

* randomize the phase angles again in (3.4.3) with F9 (or from the menu Math/Calculate Worksheet);
* change the delay spread in (3.4.4).

Next, the output signal compared with the input signal (the latter offset by the group delay for easier comparison).

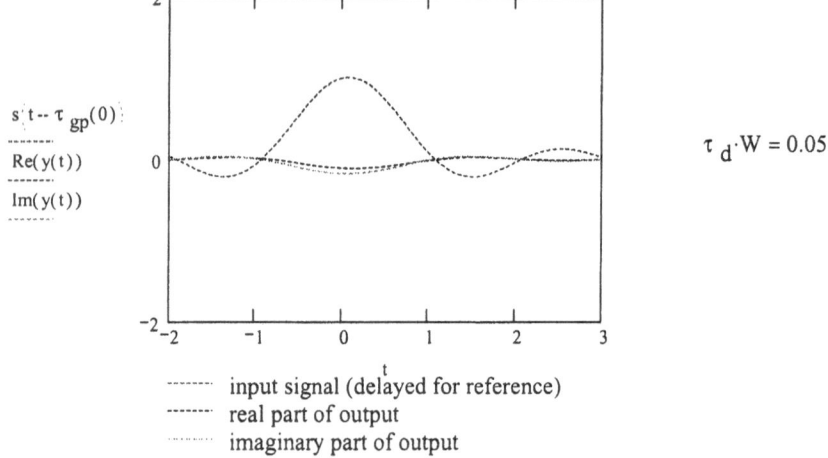

$\tau_d \cdot W = 0.05$

```
------- input signal (delayed for reference)
------- real part of output
.......... imaginary part of output
```

For small values of $\tau_d W$, the delayed reflections look much the same as the input signal, apart from gain and phase, so that the sum is almost proportional to the transmitted signal - very little distortion. For larger values of $\tau_d W$, the reflected components are far enough apart that the output signal has quite a different shape from that of the input, as evidenced in part by the difference between the real and imaginary components of the output.

4. FIRST ORDER STATISTICS OF FADING

In **Section 3**, where we first looked at fading, we saw that the received signal consists of a large number of copies of the transmitted signal, each with its own amplitude, phase and path delay. In principle, the channel is deterministic - all we have to do is measure the reflection coefficients and path lengths, and we then know how the transmitted and received signals are related. In practice, there are so many paths that this is impractical. In addition, the inevitable errors in path length measurement, even of a fraction of a wavelength, introduce phase shifts that make the resultant quite different from our prediction. In cases like this, we usually fall back on a statistical description, so that, even if we do not know the actual channel filtering, we can at least characterize some of its average properties.

This section introduces the statistical model of the channel. It discusses only the first order statistics - that is, the pdf of individual samples of, say, amplitude or instantaneous frequency. Second order statistics, such as autocorrelation functions, power spectra, joint densities, etc. are treated in **Section 5**. Even first order models, though, illustrate many of the ways our fading mobile channel differs dramatically from the additive white Gaussian noise model (AWGN) that you saw in your senior undergraduate course in communications.

4.1 Gaussian models in time and frequency

Because there are so many reflections, we use the central limit theorem to model the received signal as Gaussian. In particular, we model the complex gain $g(t)$ in flat fading as a Gaussian process. Similarly, the instantaneous frequency response $G(t_o, f)$ at some time t_o is a Gaussian process in f. If there is a line of sight (LOS) component, the processes have a nonzero mean.

Some caution is necessary, though, since very wideband signals can resolve delay differences down to the individual paths, which do not necessarily have Gaussian statistics themselves.

4.2 Rayleigh and Rice fading

With no LOS component, the complex gain is Gaussian with zero mean. In polar coordinates, its amplitude has a Rayleigh density and its phase is uniformly distributed, and the two quantities are independent. With a LOS component, giving a nonzero mean, the amplitude and phase are no longer independent. The amplitude has a Rician density and the phase pdf is rather complicated. However, both pdfs can be approximated closely by Gaussian pdfs for large values of K, the ratio of LOS power to scattered power. Another common model of amplitude variation, for both LOS and non-LOS situations, is the Nakagami density.

4.3 Consequences for BER

In static AWGN channels, we are used to an exponential decrease in bit error rate (BER) with SNR. In fading channels, however, the BER decreases only inversely with SNR. The reason is that almost all the errors occur during occasional deep fades, and the probability of such fades decreases inversely with SNR. A further consequence is that errors occur in bursts. This is serious - if we are operating at a BER of 10^{-3} on a static AWGN channel, then a 3 dB SNR improvement squares the BER, dropping it to 10^{-6}; on a fading channel, though, the same 3 dB increase would simply cut the BER in half, to 0.5×10^{-4}.

4.4 Connecting Fading, Shadowing and Path Loss

We have looked at path loss, shadowing and fading as separate phenomena. On a real link, they all affect the received signal and therefore the performance. In this section, we'll see how to combine the individual models to give an overall model suitable for system studies.

4.1 Gaussian models in time and frequency

In Section 3.2, (**3.2.7**) provided a model for the relation between transmitted and received signals. For convenience, it is reproduced here as

$$y(t) = \sum_i a_i \cdot e^{j \cdot 2 \cdot \pi \cdot f_D \cdot \cos(\theta_i) \cdot t} \cdot s(t - \tau_i) \qquad (4.1.1)$$

For narrowband signals, we have $s(t-\tau_i) \approx s(t)$, so that (4.1.1) gives the **flat fading** relation

$$y(t) = s(t) \cdot \sum_i a_i \cdot e^{j \cdot 2 \cdot \pi \cdot f_D \cdot \cos(\theta_i) \cdot t} = g(t) \cdot s(t) \qquad (4.1.2)$$

With the large number of paths, especially with their wide range of phase angles in $[-\pi, \pi)$, we can invoke the central limit theorem and model the sum g as a Gaussian random variable - and since $g(t)$ varies with time, we model it as a Gaussian random process.

This is a good time to check whether the Gaussian model is realistic. Try the **experiment in Appendix J**. It generates a large number of samples of $g(t)$ in which the the path phases θ_i are randomly drawn from a uniform distribution - the true situation - and the path amplitudes $|a_i|$ are also uniformly distributed about a mean - a reasonable, but less defensible choice - and then presents a histogram for comparison with that of a Gaussian pdf. How few arrivals you can get away with and still call the resultant reasonably Gaussian? One interesting point is that, for small to moderate numbers of arrivals, the Gaussian model overestimates the probability of deep fades. It is therefore somewhat pessimistic.

Back to the Gaussian random process model.... For a process to be Gaussian, all sets of samples, no matter how many samples in the set, must be jointly Gaussian. Clearly, that condition may not hold as the number of samples approaches the number of paths (recall again that the path amplitudes $|a_i|$ are not Gaussian themselves). However, there are usually enough paths to justify a second order (i.e., bivariate) Gaussian approximation, and that's all we need for most analyses.

Now for a different view of the scattering channel. Return to (4.1.1) and allow the signals to have wider bandwidth, but keep the channel static ($f_D=0$). This is **frequency selective transmission**. We receive

$$y(t) = \sum_i a_i \cdot s(t - \tau_i) \qquad (4.1.3)$$

The Fourier transform of the output is

$$Y(f) = S(f) \cdot \sum_i a_i \cdot e^{-j \cdot 2 \cdot \pi \cdot f \cdot \tau_i} = H(f) \cdot S(f) \qquad (4.1.4)$$

where $H(f)$ is the transfer function. Again using the central limit theorem, we can also represent the frequency response $H(f)$ as a Gaussian random process in f.

Finally, consider transmitting a tone at f through our time varying channel. From (4.1.1), the output is

$$y(t) = \sum_i a_i \cdot e^{j \cdot 2 \cdot \pi \cdot f_D \cdot \cos(\theta_i) \cdot t} \cdot e^{j \cdot 2 \cdot \pi \cdot f \cdot (t - \tau_i)}$$

$$= e^{j \cdot 2 \cdot \pi \cdot f \cdot t} \cdot \sum_i a_i \cdot e^{j \cdot 2 \cdot \pi \cdot f_D \cdot \cos(\theta_i) \cdot t} \cdot e^{-j \cdot 2 \cdot \pi \cdot f \cdot \tau_i} = e^{j \cdot 2 \cdot \pi \cdot f \cdot t} \cdot G(t, f) \qquad (4.1.5)$$

where $G(t,f)$ is the time-variant transfer function **(3.3.9)** of Section 3.3 and **(I.7)** of Appendix I. Again using the central limit theorem, we have $G(t,f)$ as a Gaussian random variable for any selection of t or f, i.e., it is a random process in time in both time and frequency. Comparison with the two cases considered earlier (Doppler spread with no delay spread, and delay spread with no Doppler spread), we see

$$g(t) = G(t, 0) \qquad H(f) = G(0, f) \qquad (4.1.6)$$

In summary:

* the complex gain $g(t)$, produced by Doppler spread, is modeled as a Gaussian process in time;

* the transfer function $H(f)$, produced by delay spread, is modeled as a Gaussian process in frequency;

* the time-variant transfer function $G(t,f)$, produced by both Doppler and delay spread, is modeled as a Gaussian process in both time and frequency.

We have to be cautious about the Gaussian assumption in some other respects, however, even if it is satisfactory for the characterizations above. For example, even signals of moderate bandwidth provide some resolution of the delay variable τ and consequently suffer some degree of frequency selectivity. If we therefore consider the time-variant impulse response $g(t,\tau)$ at any observation time t, and divide the delay axis into bins, or ranges, of width equal to the time resolution $\Delta\tau$, then

$$g(t, \text{bin\_k}) = e^{j \cdot 2 \cdot \pi \cdot f_D \cdot \cos(\theta_i) \cdot t} \sum_{i'} a_{i'} \qquad (4.1.7)$$

where i' indexes all paths with delays in bin k: $k \cdot \Delta\tau \leq \tau_{i'} < (k + 1) \cdot \Delta\tau$

As we decrease the bin width $\Delta\tau$, we have fewer and fewer paths in each bin, to the point where we cannot confidently assert that $g(t,k)$ is Gaussian. The sketch below shows an impulse response at some time t_o. It has nine paths - enough that $g(t_o)$ in (4.1.2) is Gaussian for most purposes. On the other hand, with binning, you can see from the sketch that $g(t_o, \text{bin\_1})$ and $g(t_o, \text{bin\_2})$ cannot be considered Gaussian.

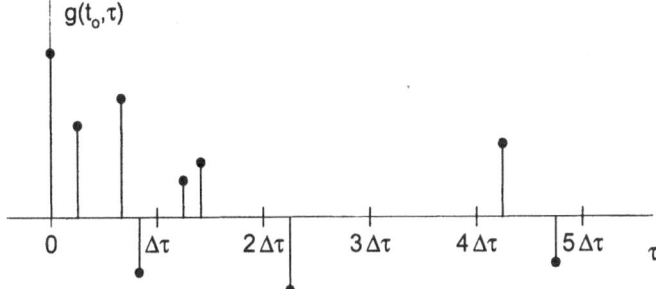

The same issue afflicts binning of the **output Doppler-spread function** $\Gamma(\nu, f)$ in Doppler shift ν for a given input frequency f, and it is even worse with the double binning of the **delay-Doppler spread function** $\gamma(\nu, \tau)$ in both ν and τ. These issues arise in the tapped delay line model for the channel, to be introduced in **Section 5.4**.

4.2 Rayleigh and Rice fading

Rayleigh Fading

From this point on, we assume that the channel complex gain or transfer function is Gaussian
For flat fading with no LOS component (i.e., zero mean), we have the variance

$$\sigma_g^{2} = \frac{1}{2} \cdot E\left[(|\,g(t)\,|)^{2}\right] = \frac{1}{2} \cdot \left[E\left[g_I(t)^2\right] + E\left[g_Q(t)^2\right]\right] \tag{4.2.1}$$

Note that the real and imaginary components are individually Gaussian with variance σ_g^2. The
probability density function is

$$p_g(g) = \frac{1}{2 \cdot \pi \cdot \sigma_g^{2}} \cdot \exp\left[-\frac{1}{2} \cdot \frac{(|\,g\,|)^2}{\sigma_g^{2}}\right] \tag{4.2.2}$$

and its isoprobability contours are circles centred on the origin:

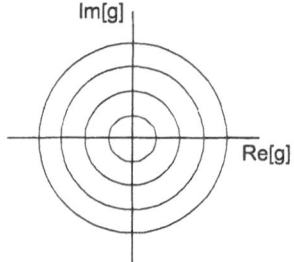

If we change to polar coordinates $g = g_I + j \cdot g_Q = r \cdot e^{j \cdot \theta}$ then standard transformations
[**Papo84, Proa95, Lee82**] give the joint pdf as

$$p_{r\theta}(r,\theta) = \frac{r}{2 \cdot \pi \cdot \sigma_g^{2}} \cdot \exp\left(-\frac{r^2}{2 \cdot \sigma_g^{2}}\right) \tag{4.2.3}$$

Clearly, r and θ are independent, since the joint pdf is the product of their individual pdfs, given
by

$$p_r(r, \sigma_g) := \frac{r}{\sigma_g^{2}} \cdot \exp\left(-\frac{r^2}{2 \cdot \sigma_g^{2}}\right) \quad r \geq 0 \quad \text{and} \quad p_\theta(\theta) = \frac{1}{2 \cdot \pi} \quad -\pi \leq \theta < \pi \tag{4.2.4}$$

This pdf of the amplitude r is the *Rayleigh distribution*, and this type of fading (no LOS component) is termed Rayleigh fading. Let's see what it looks like.

$$i := 0..100 \qquad r_i := \frac{5}{100} \cdot i \qquad \text{Your choice of } \sigma_g: \quad \sigma_g := 1$$

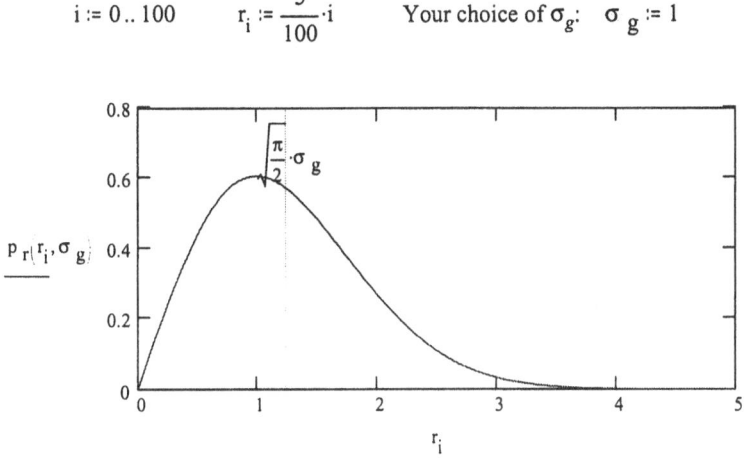

Rayleigh distribution of amplitude

mean   mode   standard deviation

$$\mu_r = \sqrt{\frac{\pi}{2}} \cdot \sigma_g \qquad\qquad \sigma_g \qquad\qquad \sigma_r = \sqrt{2 - \frac{\pi}{2}} \cdot \sigma_g$$

It's often easier to work with the squared amplitude (twice as large as instantaneous power)

$$z = r^2 = (|g|)^2$$

because it is exponentially distributed. This follows from a simple change of variables to z from r in the Rayleigh pdf. Alternatively, note that $z = g_I^2 + g_Q^2$, and that the sum of independent squared Gaussian variates has the χ^2 distribution, and that the χ^2 distribution with two degrees of freedom is exponential. In any case, the pdf of z is

$$p_z(z, \sigma_g) := \frac{1}{2 \cdot \sigma_g^2} \cdot e^{-\frac{z}{2 \cdot \sigma_g^2}} \qquad z \geq 0 \qquad\qquad (4.2.5)$$

Again, plot it:

$$z_i := (r_i)^2 \qquad\qquad \text{Your choice of } \sigma_g: \quad \sigma_g := 1$$

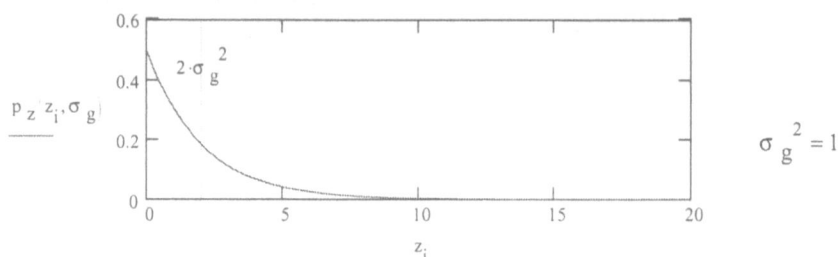

Exponentially distributed squared magnitude

The mean equals the decay constant   and so does the standard deviation

$$\mu_z = 2 \cdot \sigma_g^2$$

$$\sigma_z = 2 \cdot \sigma_g^2$$

The cumulative distribution function of z and its asymptote are

$$F_z\left(z, \sigma_g\right) := 1 - \exp\left(-\frac{z}{2 \cdot \sigma_g^2}\right) \qquad \text{aympt}\left(z, \sigma_g\right) := \frac{z}{2 \cdot \sigma_g^2} \qquad (4.2.6)$$

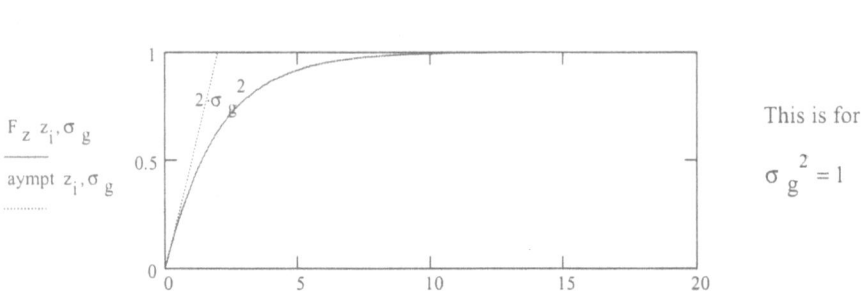

The asymptote gives us two very useful rules of thumb. Remember them:

* **The probability that received power is 10 dB or more below the mean level (a 10 dB fade) is 10%; probability of a 20 dB fade is 1%; probability of a 30 dB fade is 0.1%; etc.** Now go back to the fade graph in __Section 3.1__ and see whether this seems to be true (recalculate a few times). Remember that with $\lambda/50$ sampling, some deep fades may be missed, so you are really looking at the fraction of the number of samples M that falls below the threshold.

* **The probability that the power drops below a given level decreases only inversely with increasing average power σ_g^2.** That's important - and disappointing - if the level is a threshold below which operation is unacceptable, since doubling the average power only cuts the probability in half!

Rice Fading

In mobile satellite systems, or in land mobile radio in suburban and rural areas, the signal is often received with a LOS component which produces Rice fading. The total gain

$$g = g_s + g_d \tag{4.2.7}$$

is the sum of a constant specular (or LOS or discrete) component g_s and a zero mean Gaussian diffuse (or scattered) component g_d, so that g is a nonzero mean Gaussian variate. The specular component has K times the power of the diffuse component (the Rice K-factor), so that $K=0$ gives Rayleigh fading and $K ==> \infty$ gives a constant channel. But be careful - some literature (mostly in the mobile satellite area) uses K as the ratio of diffuse to specular power, the reciprocal of the conventional definition. The sketch shows the isoprobability contours.

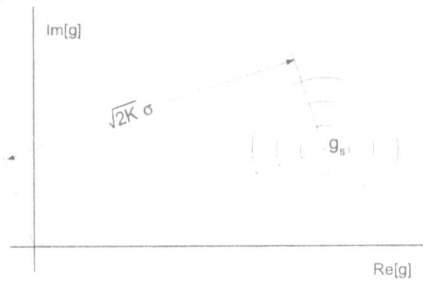

Denote the variance of the diffuse component by σ^2. From the power ratio we have the magnitude of the specular component.

$$\frac{1}{2} \cdot E\left[\left(|g_d|\right)^2\right] = \sigma^2 \qquad\qquad |g_s| = \sqrt{2 \cdot K} \cdot \sigma \tag{4.2.8}$$

The total average power in g is then

$$\frac{1}{2} \cdot E\left[\left(|g|\right)^2\right] = \frac{1}{2} \cdot E\left[\left(|g_s|\right)^2\right] + \frac{1}{2} \cdot E\left[\left(|g_d|\right)^2\right] = K \cdot \sigma^2 + \sigma^2 = \sigma^2 \cdot (1 + K) \tag{4.2.9}$$

and the mean and variance of g are $\mu_g = g_s$ and $\sigma_g^2 = \sigma^2$. Its pdf is Gaussian:

$$P_g(g) = \frac{1}{2 \cdot \pi \cdot \sigma_g^2} \cdot \exp\left[-\frac{1}{2} \cdot \frac{\left(|g - \mu_g|\right)^2}{\sigma_g^2}\right] \tag{4.2.10}$$

Changing to polar coordinates makes $z = r^2$ non-central χ^2 with mean $\sigma^2(1+K)$ and two degrees of freedom. Alternatively, the pdf of r is Rician:

$$p_{r\_K}(r,K,\sigma) := \frac{r}{\sigma^2} \cdot \exp\left(-\frac{r^2}{2 \cdot \sigma^2} - K\right) \cdot I0\left(\frac{r \cdot \sqrt{2 \cdot K}}{\sigma}\right) \tag{4.2.11}$$

From the isoprobability sketch above, it is clear that the phase angle is not independent of the amplitude. The unconditional pdf of the phase angle for a real specular component (i.e., zero mean phase angle) is obtained by adapting [**Proa89**, eqn. 4.2.103]. First, the Q function:

$$Q(x) = \frac{1}{\sqrt{2 \cdot \pi}} \cdot \int_x^\infty e^{-\frac{\alpha^2}{2}} d\alpha \qquad \text{or} \quad Q(x) := cnorm(-x)$$

$$p_{\theta\_K}(\theta,K,\sigma) := \frac{1}{2 \cdot \pi} \cdot e^{-K} \cdot \left[1 + \sqrt{4 \cdot \pi \cdot K} \cdot \cos(\theta) \cdot e^{K \cdot \cos(\theta)^2} \cdot \left(1 - Q\left(\sqrt{2 \cdot K} \cdot \cos(\theta)\right)\right)\right]$$

$$\tag{4.2.12}$$

Now let's see what these pdfs look like.

plot ranges: $r := 0, 0.05 .. 10$ $\theta := -\pi, -0.98 \cdot \pi .. \pi$

Your choice of K and σ below

$K := 2$ $\sigma := 1$

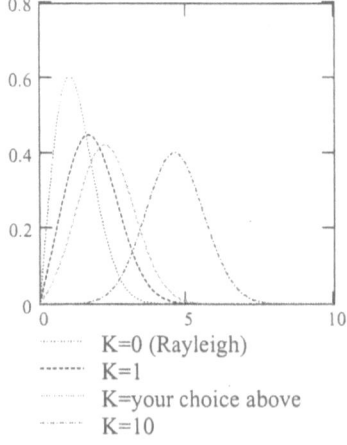

- K=0 (Rayleigh)
- K=1
- K=your choice above
- K=10

Rice amplitude pdf

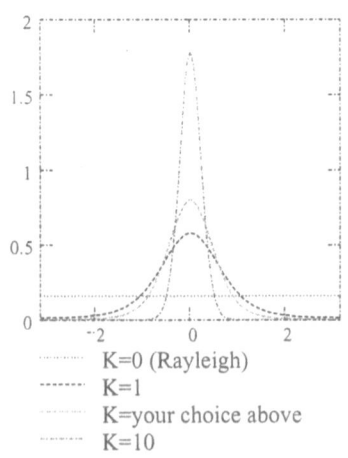

- K=0 (Rayleigh)
- K=1
- K=your choice above
- K=10

Rice phase pdf (for zero mean)

Inspection of the graphs suggests that they can be approximated by Gaussian pdfs for large K. It's easy to see why if you rotate the coordinates for g_d to resolve it into a radial component (along the same line as g_s) and a transverse component (orthogonal to the radial component). For large K, the transverse component makes little difference to the amplitude, which is then well modeled by the Gaussian radial component with the specular component as a mean. Similarly, the radial component makes little difference to the phase, which is then well modeled by the Gaussian transverse component divided by the specular amplitude. Therefore,

* approx amplitude pdf, large K: Gaussian, mean $\sqrt{2 \cdot K} \cdot \sigma$ and standard deviation σ

* approx phase pdf, large K: Gaussian, mean $\arg\left(g_s\right)$ and standard deviation $\dfrac{1}{\sqrt{2 \cdot K}}$

Nakagami Density

When you did the **experiment in Appendix J**, you noticed that the Rayleigh pdf was a fairly rough fit to the histogram of experimental values if the number of paths was small. Many authors report that a better approximation of their experimental measurements is obtained with the Nakagami-m distribution, given by [**Naka60**]

$$p_{r\_N}(r,m,\sigma) := \frac{2}{\Gamma(m)} \cdot \left(\frac{m}{2 \cdot \sigma^2}\right)^m \cdot r^{2 \cdot m - 1} \cdot e^{\frac{-m \cdot r^2}{2 \cdot \sigma^2}} \qquad \text{for } r \geq 0 \quad \text{and} \quad m \geq \frac{1}{2} \qquad (4.2.13)$$

where m is the order of the pdf, $2\sigma^2$ is the mean square value and $\Gamma(m)$ is the gamma function (equal to $(m-1)!$ for integers). For $m=1$, Nakagami reduces to Rayleigh. Let's have a look at it for $\sigma := 1$:

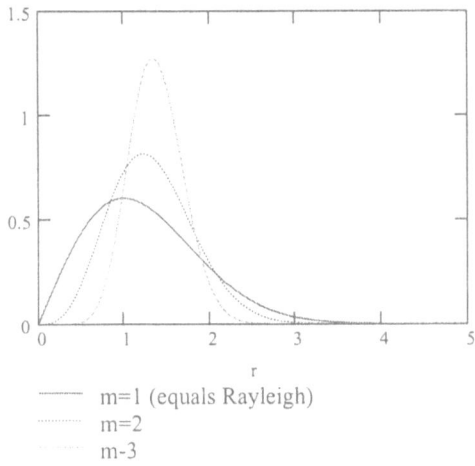

——— m=1 (equals Rayleigh)
.... m=2
.... m-3

The Nakagami-m pdf

You can see that increasing the order m of the Nakagami distribution changes its character from that of purely scattered fading to fading with a LOS component. For modeling these channels, it is therefore a reasonable alternative to the Rice pdf which it resembles. For larger values of m, just as for larger values of K in the Rice pdf, it can be approximated in turn by a Gaussian pdf.

Why bother with this new pdf, when we already have the Rice pdf? One reason is its simplicity. For example, by change of variables, the squared amplitude $z=r^2$ has a gamma pdf:

$$P_{z\_N}(z,m,\sigma) := \frac{1}{\Gamma(m)} \cdot \left(\frac{m}{2 \cdot \sigma^2}\right)^m \cdot z^{m-1} \cdot e^{\frac{-m \cdot z}{2 \cdot \sigma^2}} \qquad \text{for} \quad z \geq 0 \quad \text{and} \quad m \geq \frac{1}{2} \qquad (4.2.14)$$

This looks more complicated than it is - just focus on the variation with z and it looks like functions you have seen before in your undergraduate course on linear systems and Laplace transforms. Consequently, its characteristic function (Laplace transform of pdf) is

$$M_{z\_N}(s,m,\sigma) := \left(\frac{m}{2 \cdot \sigma^2} \cdot \frac{1}{s + \frac{m}{2 \cdot \sigma^2}}\right)^m \qquad (4.2.15)$$

which has an mth order pole at $s = \frac{-m}{2 \cdot \sigma^2}$ You can obtain many analytical results conveniently with these expressions, in contrast to the Rice pdf (4.2.11), with its embedded Bessel function. For representative work using the Nakagami pdf see [**Pate97**], [**Ugwe97**].

4.3 Consequences for BER

This section demonstrates a dramatic difference in performance between fading channels and static AWGN channels. We will see that the behaviour is determined almost exclusively by deep fades.

The question: what is the expected BER if we place a mobile in a randomly selected position? Equivalently, what is the average BER as the mobile moves through the random field? To simplify the discussion, assume:

* flat fading, so that the channel does not introduce intersymbol interference;
* slow fading, so that the primary effect is an SNR that varies slowly compared to the bit rate; i.e., we can ignore the pulse distortion and phase hits that occur in deep fades;
* Rayleigh, not Rice, fading.

Obviously, the BER depends on the received signal amplitude r, so we write it $P_e(r)$. Then the quantity of interest is

$$P_{e\_av} = E_r(P_e(r))   \tag{4.3.1}$$

where $E_r(\ )$ denotes expectation with respect to r. Because the $P_e(r)$ function is very nonlinear, the average $P_{e\_av} \neq Pe(r_{\_av})$, and we will have to compute (4.3.1) explicitly. Here it's easier to work with squared amplitude, because it has a simple exponential distribution. Denote the instantaneous SNR as $\gamma_b = E_b/N_o$. It is proportional to the average SNR Γ_b times z:

$$\gamma_b = \Gamma_b \cdot z   \qquad \text{where} \qquad z = r^2 \quad \text{and} \quad E_r(z) = 1 \quad ==> \quad \sigma_g^2 = \frac{1}{2}   \tag{4.3.2}$$

so that $\quad \Gamma_b = E_r(\gamma_b)$

Again for simplicity, assume that we transmit orthogonal signals and detect them incoherently. Then the BER as a function of z is

$$P_e(z, \Gamma_b) := \frac{1}{2} \cdot e^{-\frac{\Gamma_b \cdot z}{2}}   \tag{4.3.3}$$

and the average BER is obtained by averaging with respect to the exponential distribution of z:

$$P_{e\_av}(\Gamma_b) = \int_0^\infty P_e(z, \Gamma_b) \cdot e^{-z} dz = \frac{1}{2} \cdot \int_0^\infty e^{-\frac{\Gamma_b \cdot z}{2}} \cdot e^{-z} dz   \tag{4.3.4}$$

giving

$$P_{e\_av}(\Gamma_b) := \frac{1}{2 + \Gamma_b}   \qquad \text{which is asymptotic to } \Gamma_b^{-1}   \tag{4.3.5}$$

Let's compare static channel (4.3.3) and fading channel (4.3.5) BERs with the same average SNR.

> ⊡ Reference:D:\COURSES\MobChann\paperbook\Units.mcd(R)

$$P_{e\_stat}\left(\Gamma_b\right) := larger\left(P_e\left(1,\Gamma_b\right),10^{-14}\right)$$ (to prevent log scale on graph from bottoming out)

$$i := 0..40 \qquad \Gamma_{bdB_i} := i \qquad \Gamma_{b_i} := nat\left(\Gamma_{bdB_i}\right) \qquad (nat = inverse\ dB)$$

SNR per bit, dB
.... static channel
.... Rayleigh fading channel

We see a major difference in behaviour - at 10^{-4}, there is roughly 27 dB between the curves! Putting it another way:

* On a static channel, a 3 dB increase in SNR *squares* the BER (e.g., from 10^{-3} to 10^{-6}).

* On a fading channel, a 3 dB increase in SNR only *halves* BER (e.g., from 10^{-3} to 0.5×10^{-3})

Why is there such a drastic difference between static and fading channels? The reason is straightforward:

* $P_e(x)$ is strongly nonlinear, so BER at low SNR is *much* greater than at high SNR.

* When averaging numbers (e.g., BERs) with large dynamic range, such as

$$10^{-6}, 10^{-7}, 10^{-5}, 10^{-6}, 10^{-2}, 10^{-5}, 10^{-6}$$

the sum is determined primarily by the very large values and their probability of occurrence.

So deep fades dominate the behaviour, and the integrand in $\int_0^\infty P_e(x) \cdot p_x(x) \, dx$ is significant only for values of x much less than the mean. How likely are those low values? Remember the rule of thumb from **Section 4.2**: from the asymptote for Rayleigh fading, the probability of being below a specific SNR level decreases only inversely with increasing average power. That's why BER decreases as $1/\Gamma_b$ on Rayleigh channels, instead of exponentially, as it does on static channels.

Here's another important consequence of the dominance of deep fades: error occur in bursts. In mobile communications, we have long intervals - in fact, most of the time - with no errors at all, separated by clusters of several errors. You can see why from the plots of signal strength against time (or position) below. We'll use the Jakes complex gain generator and binary incoherent signaling.

→ Reference:D:\COURSES\MobChann\paperbook\Jakesgen.mcd(R)

select M points spaced by 0.02λ:

$\lambda := 1$ $M := 200$ $i := 0 .. M$ $x_i := 0.02 \cdot \lambda \cdot i$

$G := \text{Jakes\_init}(0)$ $g_i := \text{Jakes\_gen}(x_i, G)$ this gives variance $\lambda/2$, as required

$\Gamma_{bdB} := 25$ in dB $\Gamma_b := \text{nat}(\Gamma_{bdB})$ converts from dB

$p_i := P_{e\_stat}\left[\Gamma_b \cdot \left(|g_i|\right)^2\right]$ see definition before previous graph

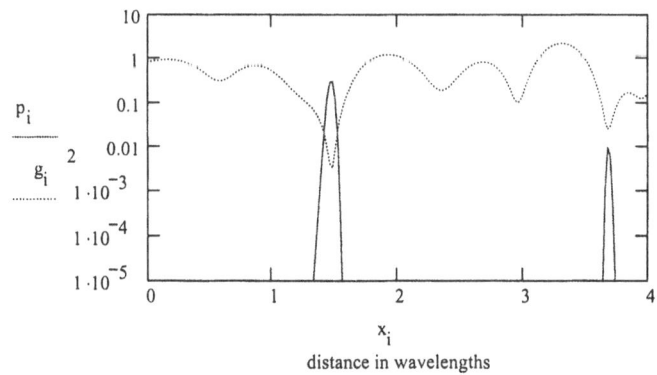

To see other complex gain patterns, recalculate the Jakes initialization with F9.

x_i

distance in wavelengths

——— BER

············ magnitude squared of complex gain

BER as a Function of Position (Time)

This is a striking illustration of the burstiness of error patterns in mobile communications.

4.4 Connecting Fading, Shadowing and Path Loss

Some Observations

So far, you have studied path loss, shadowing and fading separately. On a real link, though, they are all present, and you must account for all of them in any system design. In this section, you'll see how they link up. You'll also be introduced to two generic ways of analyzing the resulting error rate.

First, recall that **path loss** describes the tendency of signal power to decrease as an inverse power of distance from the transmitter. Typically, it is taken as inverse fourth power, so that

$$P_p = \frac{P_0}{r^4} \qquad\qquad (4.4.1)$$

where the subscript p denotes path loss. Taken alone, (4.4.1) implies that the received signal power (at a mobile from the base, or at the base from a mobile) is predictable and constant at all points in a thin annulus at a distance r from the base. As the sketch below reminds us, though, this is far from the case.

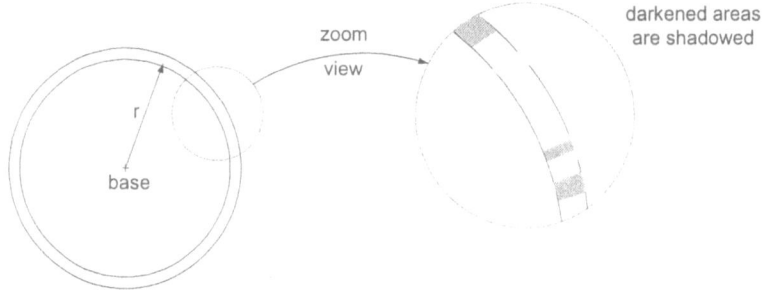

The sketch illustrates much variation within the annulus, caused by **shadowing**. Consequently, we interpret (4.4.1) as an average over all points in the annulus and, in fact, over a large number of typical cells. The scale of shadowing variations depends on the size of the obstacles between the base and mobile, but we usually expect it to be roughly constant over a few tens of wavelengths and to exhibit significant variation over hundreds of wavelengths. The variation is usually **modeled statistically**, rather than calculated deterministically, with a **lognormal distribution**. The received signal power, after accounting for both path loss and shadowing, was given as **(2.1.1)**, and is repeated here as

$$P_{ps} = \frac{P_0}{r^4} \cdot 10^{0.1 \cdot z} \qquad\qquad (4.4.2)$$

where z is a Gaussian random variable with zero mean and standard deviation in the range 6 dB to 10 dB, depending on the terrain, and the subscript ps denotes combined path loss and shadowing.

Taken alone, (4.4.2) suggests that signal power is constant over a patch a few tens of wavelengths in diameter. However, we know it is not - **fading**, caused by interference among scattered components at the mobile, imposes a local fine structure with a half-wavelength scale of variation. The sketch below shows the variations due to all three phenomena: path loss, shadowing and fading.

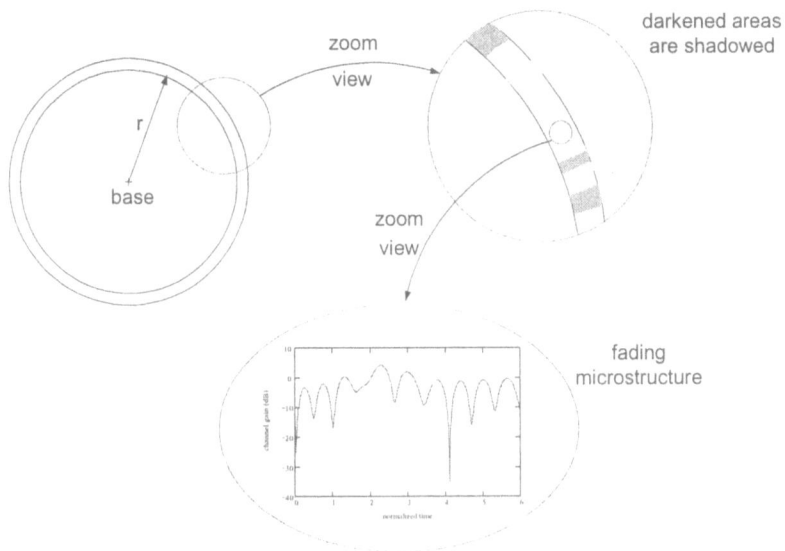

Fading is a multiplicative phenomenon that is also modeled statistically, usually by a **Rayleigh or Rice** pdf, but the **Nakagami** pdf is also sometimes used. Consequently, we can identify the mean square value of the received signal over a patch a few tens of wavelengths in diameter with a sample value from the power variation due to fading and shadowing (4.4.2), which operates on a larger spatial scale.

To describe the received signal $r(t)$, it is convenient to normalize the transmitted signal $s(t)$ to unit second moment. In flat fading, the complex gain $g(t)$ can also be normalized to unit second moment. That is, $E\left[(|\ s\ |)^2\right]=1$ and $E\left[(|\ g(t)\ |)^2\right]=2\cdot\sigma_g^2=1$ The received signal is then

$$r(t)=\sqrt{2\cdot P_{ps}}\cdot g(t)\cdot s(t)=\sqrt{2\cdot P_p\cdot 10^{0.1\cdot z}}\cdot g(t)\cdot s(t)=\sqrt{2\cdot P_o\cdot\frac{10^{0.05\cdot z}}{r^2}}\cdot g(t)\cdot s(t) \qquad (4.4.3)$$

and the power in $r(t)$, conditioned on the shadowing and fading values, is therefore

$$P_{psf}=\frac{1}{2}\cdot E_s\left[(|\ r(t)\ |)^2\right]=\frac{P_o}{r^4}\cdot 10^{0.1\cdot z}\cdot(|\ g\ |)^2=\frac{P_o}{r^4}\cdot 10^{0.1\cdot z}\cdot x \qquad (4.4.4)$$

where x, the squared magnitude of the fading gain g, is exponentially distributed with unit mean since it is Rayleigh. If we average over the fading, we have the received power as a function of path loss and shadowing

$$P_{psf} = \frac{P_0}{r^4} \cdot 10^{0.1 \cdot z} \tag{4.4.5}$$

which, of course, equals P_{ps} (4.4.2). For **frequency selective fading**, we have a convolution relationship between transmitted and received signals, and we take the sum of gain second moments on all paths as unity.

Analysis by Conditional Fade pdf

Channel fading produces transmission errors, not just by its occasional deep fades and low SNR, but also by the rapid changes in amplitude and phase. As you will see shortly, the distortion produced by fast fading results in a few errors even if there is no additive noise at all.

This suggests that a good way to analyze the channel behaviour is to fix the value of the shadowing random variable, which also fixes the second moment and SNR of the received signal. Focus your analysis on the effect of fading. The resulting bit error rate (or other measure) is really a conditional mean, conditioned on a specific value of shadowing. To complete the statistical part of the analysis, average the result over the SNR pdf, which is lognormal because of shadowing. This second average produces a deterministic function of the path distance, to which you can apply path loss models for mean SNR. Most analyses are in fact conducted in this way.

Analysis by Combined Fade pdf

An alternative approach is possible in slow fading. If the fading is slow enough that signal distortion is negligible, and the dominant cause of errors is just the intervals of low SNR, all you need is the pdf of signal amplitude or signal power. In this case, you might consider combining the exponential pdf of fading power in flat Rayleigh fading with the lognormal pdf of shadowing power. This approach was pursued in [**Fren79**] and other related results are summarized in [**Stub96**].

This approach is not as promising as it sounds. Start by expressing the pdf of received power, conditioned on the shadowing and the fading. From (4.4.4),

$$P_{psf} = P_p \cdot w \cdot x = P_p \cdot u \tag{4.4.6}$$

where the shadowing value is denoted here by $w = e^y$ and the fading value by $x = |g|^2$. Their product, the composite variable for which we seek the pdf, is denoted by u. Since w and x are independent with unit means, u also has unit mean. The pdf of u, conditioned on w, follows from the fact that x has a unit mean exponential pdf:

$$p_{u\_w}(u \mid w) = \frac{1}{w} \cdot e^{-\frac{u}{w}} \tag{4.4.7}$$

Consequently, the unconditional pdf of u is obtained using the lognormal pdf of w as

$$p_u(u) = \int_0^\infty \frac{1}{\sqrt{2 \cdot \pi} \cdot \sigma_y \cdot w^2} \cdot e^{-\frac{u}{w}} \cdot \exp\left[-\frac{1}{2} \cdot \left(\frac{\ln(w) - \mu_y}{\sigma_y} \right)^2 \right] dw \tag{4.4.8}$$

Finally, recall from **Appendix A** that a unit mean for w requires that the mean of its logarithm satisfy $\mu_y = -\sigma_y^2/2$. Substitution into (4.4.8) and simplification gives

$$p_u(u) = \int_0^\infty \frac{1}{\sqrt{2 \cdot \pi} \cdot \sigma_y \cdot w^2} \cdot \exp\left[-\frac{u}{w} - \frac{1}{2} \cdot \left(\frac{\ln(w)}{\sigma_y} + \frac{\sigma_y}{2} \right)^2 \right] dw \tag{4.4.9}$$

This is the desired composite pdf of received power. Its counterpart for received amplitude, obtainable by a simple change of variable, is known as the Suzuki distribution. However, the integral does not have a closed form, and must be evaluated numerically or approximated. This feature limits its value in analysis, which is why I characterized it above as not as being promising as it seems.

5. SECOND ORDER STATISTICS OF FADING

In **Section 4.1**, we concluded that, given enough scatterers, we can reasonably model the complex gain process $g(t)$ and the instantaneous frequency response $H(f)$ by Gaussian processes. As you know, to characterize a Gaussian process, it is necessary and sufficient to specify its second order statistics: the autocorrelation function or its transform, the power spectrum. That's the subject of this section.

Some of the models for second order behaviour that we'll see are simplified or idealized. Nevertheless, they are widely used in design and in the literature because they provide a well-understood benchmark that facilitates performance estimation and comparison of alternative methods.

In these subsections, we'll develop the Doppler spectrum and the delay power profile as second order descriptions of **time-selective and frequency-selective fading**, respectively, and we will also (briefly) examine the more difficult case of fast frequency-selective fading in which Doppler and delay spread are not separable. They have a happy ending, though, in the form of the very useful TDL time-variant channel model that you can use with confidence for most situations.

There are also two appendices that deal with joint statistics of a process and its derivative, and similar statistics expressed in polars. They may seem a little esoteric, but they will arise occasionally in your work, and you need some of the results in order to understand level crossing rates in **Section 6.2**.

5.1 Doppler Spectrum and WSS Channels

What are the power spectrum and autocorrelation function of the complex gain process $g(t)$? We'll consider scatterers around the mobile and account for the fact that their Doppler shift depends on their angle with respect to the direction of movement. The special case of isotropic scattering, with a uniform density of scatterers in azimuth around the mobile, is almost universally employed as a benchmark model for fading at the mobile.

5.2 Power Delay Profile and US Channels

The correlation between signals received on carriers at different frequencies is important in the design of frequency diversity systems, CDMA systems and even multicarrier OFDM systems. We see how it is related to the power delay profile and common measures like rms delay spread.

5.3 Scattering Function and WSSUS Channels

In frequency selective fading, considerable simplification results if the scatterers at different delays or Dopplers are uncorrelated. This is the commonly used WSSUS (wide sense stationary, uncorrelated scattering) model, summarized by the scattering function, which is introduced in this section. Even in the WSSUS case, the distribution of the scatterers may cause delay spread and Doppler spread to be linked, so that the power spectrum of the complex gain may depend on the delay of the scatterers involved. Usually, we try to ignore it.

5.4 A TDL Channel Model for Analysis and Simulation

For signals with bandwidths large enough that we must consider delay spread, yet not so large that they resolve the delay differences of individual paths, we have another near-universal model of frequency-selective fading. It is just a tapped delay line (TDL) with coefficients that vary with time according to their individual complex gain processes. This one is very useful in simulation work, as well as analysis.

Appendix F: Joint Second Order Statistics

We are often interested in the dynamics, or time variations, of fading - for example, in computing error floors, or estimating the effect on error correcting codes. That leads us to the statistics of the derivative of the complex gain process, and its correlation with the complex gain itself. Both processes are Gaussian, so joint pdfs are easily obtained.

This material is here primarily as a reference. Skim through it to see what it contains, so that you can retrieve the material when you need it.

Appendix G: Joint pdfs in Polar Coordinates

When we convert to polar coordinates, the pdfs relating the complex gain and its derivative become difficult. Many of them exist in closed form, though, and are useful. As with Appendix F, the material is here as a reference in case you need it. We'll refer to some of it when discussing random FM and level crossing rates.

5.1 Doppler Spectrum and WSS Channels

We saw in **Section 3.3** that the input and output of a flat fading channel are related by $y(t)=g(t)s(t)$ where $g(t)$ is the complex gain. In particular, if we transmit the unit amplitude unmodulated carrier, $s(t)=1$, then we observe $y(t)=g(t)$. A spectrum analyzer reveals that the tone is broadened by the Doppler shifts.

In this section, we derive the power spectrum and the autocorrelation of the complex gain. These functions are exceedingly important in analysis of modulation on fading channels - in calculation of pulse distortion effects, error floors, effectiveness of interleaving, and many, many other situations.

Power Spectrum

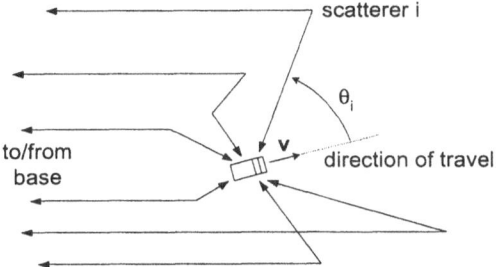

Start by considering the sketch of a mobile in a cluster of scatterers. It is moving at speed v, and scatterer i is at an angle θ_i with respect to the direction of motion. As we have previously observed, the Doppler shift ranges from $f_D=v/\lambda$ caused by reflections directly ahead, to $-f_D$, caused by reflections directly behind. In general, the Doppler shift v (not the same symbol as speed v) of a reflector is

$$v = f_D \cdot \cos(\theta) \tag{5.1.1}$$

Note that, since $\cos(\theta)$ is an even function, a Doppler shift of v comes from either side of the direction of motion, $\theta = \pm\arccos(v/f_D)$, as illustrated by the sketch below.

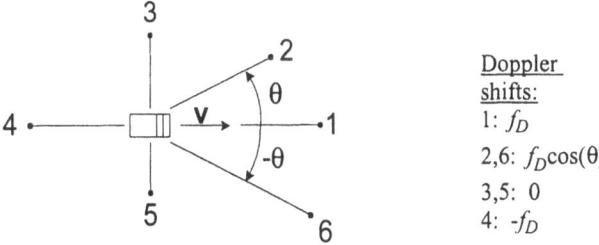

Next we relate differential ranges of Doppler shift frequency v and angle θ by differentiating (5.1.1). We have

$$dv = -f_D \cdot \sin(\theta) \cdot d\theta = -f_D \cdot \sqrt{1 - \cos(\theta)^2} \cdot d\theta = -f_D \cdot \sqrt{1 - \left(\frac{v}{f_D}\right)^2} \cdot d\theta \qquad (5.1.2)$$

For convenience, assume that there are enough scatterers that we can work with densities. Denote the incoming signal power in differential angle $d\theta$ at θ by $P(\theta)d\theta$, and denote the antenna amplitude gain pattern by $A(\theta)$, both referenced to the direction of motion, defined by $\theta=0$.

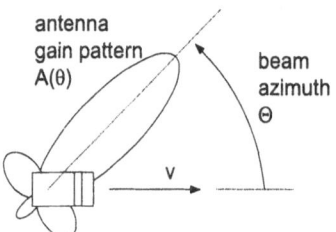

Then the received power arriving in $d\theta$ at θ is

$$A(\theta)^2 \cdot P(\theta) \cdot d\theta \qquad (5.1.3)$$

and the received power in differential frequency dv at v is

$$S_g(v) \cdot dv = \left(A(\theta)^2 \cdot P(\theta) + A(-\theta)^2 \cdot P(-\theta) \right) \cdot \frac{dv}{f_D \cdot \sqrt{1 - \left(\frac{v}{f_D}\right)^2}} \qquad (5.1.4)$$

where, of course, $\theta = arccos(v/f_D)$. We have just obtained the power density spectrum of the complex gain - that is, the Doppler spectrum - as $S_g(v)$. The generality of (5.1.4) will be very useful when we consider directionality of antennas at the mobile and base station in **Section 5**.

At this point, we'll consider the very important special case of isotropic scattering (i.e., $P(\theta)$ is uniform) and an isotropic antenna, such as a vertical whip (so $A(\theta)$ is uniform). Then the Doppler spectrum

$$S_g(v) \text{ is proportional to } \frac{1}{\sqrt{1 - \left(\frac{v}{f_D}\right)^2}}$$

and normalizing so that the total power is σ_g^2, we have

$$S_g(v) = \frac{\sigma_g^2}{\pi \cdot f_D} \cdot \frac{1}{\sqrt{1 - \left(\frac{v}{f_D}\right)^2}}$$

(5.1.5)

This is the well-known U-shaped spectrum. With loop antennas, we have an antenna directionality $A(\theta)$ of $\cos(\theta)$ or $\sin(\theta)$, depending on orientation, with easily calculable results. To plot it, normalize the frequency $u = v/f_D$ and define

$$\text{whip}(u) := \frac{1}{\sqrt{1 - u^2}} \qquad \text{loop1}(u) := \frac{u^2}{\sqrt{1 - u^2}} \qquad \text{loop2}(u) := \sqrt{1 - u^2} \qquad (5.1.6)$$

The plot range is defined in a way that avoids singularities

$$u := -0.99, -0.98 .. 0.99$$

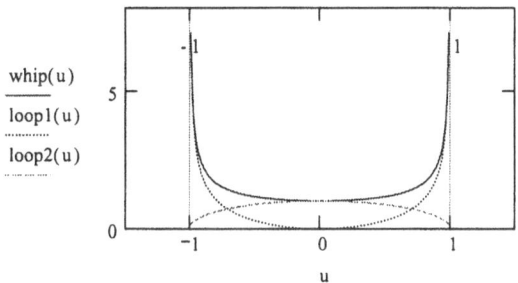

Normalized Frequency u=nu/fD

Doppler Spectrum Sg(nu)

The U-shaped spectrum of isotropic scattering is ubiquitous in analyses of signal detection in mobile channels. But is it realistic? For pedestrian use, it's not bad. For vehicular applications, it has some flaws. For one, how often do cars have large scatterers directly ahead or behind? They are the ones that produce the $\pm f_D$ shifts, so that in reality, the spectrum is flatter than the one we derived. Another point is that the environment is not motionless - cars in the lane facing us also act as scatterers, and therefore contribute some frequency shifts up to $\pm 2 f_D$.

In addition to those considerations, there's the fact that, in many situations, the environment around the mobile could be dominated by a few large scatterers, rather than a collection of uniformly distributed small scatterers. Each large scatterer or cluster of scatterers produces a peak in the power spectrum at a Doppler shift v corresponding to its azimuth θ. The sketch below shows such a physical situation, with reflectors of various sizes and locations (identified by labels 1 to 4). Beneath it is the corresponding Doppler spectrum - which is not even symmetric.

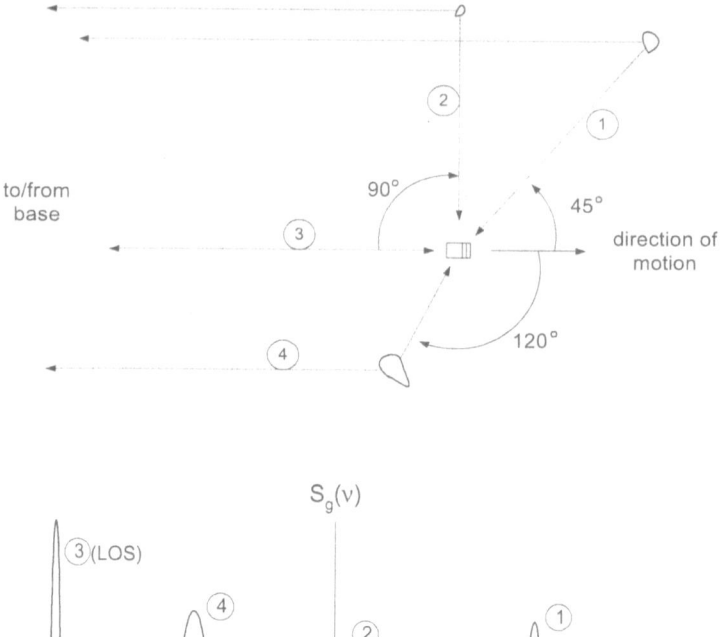

From these observations, you can see that the U-shaped spectrum derived from isotropic scattering is very idealized. Nevertheless, it's not a bad testbed for comparison of different systems and techniques, because it approximates an ensemble average of $S_g(v)$ taken over a large number of locations.

The rms Doppler spread is obtained from the spectrum like any rms bandwidth. We recognize that $S_g(v)$ is positive, then normalize it to unit area. The mean frequency is then

$$v_m = \int_{-\infty}^{\infty} v \cdot S_g(v) \, dv \cdot \left(\int_{-\infty}^{\infty} S_g(v) \, dv \right)^{-1}$$

$$= \frac{1}{\sigma_g^2} \cdot \int_{-\infty}^{\infty} v \cdot S_g(v) \, dv \tag{5.1.7a}$$

where the second line follows from the normalization in (5.1.5) to make the area under the power spectrum equal to σ_g^2. Next, the mean square bandwidth (squared rms bandwidth) is the central moment

$$v_{rms} = \frac{1}{\sigma_g^2} \cdot \int_{-\infty}^{\infty} (v - v_m)^2 \cdot S_g(v) \, dv \qquad (5.1.7b)$$

In the *special case* that the fading spectrum is symmetric, with any frequency offset accounted for separately in the analysis, (5.1.7b) reduces to

$$v_{rms}^2 = \frac{1}{\sigma_g^2} \cdot \int_0^{f_D} v^2 \cdot S_g(v) \, dv \qquad (5.1.7c)$$

Note the similarity with calculation of means and variances - pdfs are also nonnegative with unit area.

Substitution of (5.1.5) into (5.1.7c) yields (after a change of variable)

$$v_{rms} = \frac{f_D}{\sqrt{2}} \qquad (5.1.8)$$

Autocorrelation Function

Just as important is the autocorrelation function of the complex gain. We know that it is the inverse Fourier transform of the power spectrum, so for observation times separated by t

$$R_g(\tau) = \int_{-f_D}^{f_D} \frac{\sigma_g^2}{\pi \cdot f_D} \cdot \frac{1}{\sqrt{1 - \left(\frac{v}{f_D}\right)^2}} \cdot e^{j \cdot 2 \cdot \pi \cdot v \cdot \tau} \, dv = \frac{\sigma_g^2}{2 \cdot \pi} \cdot \int_{-\pi}^{\pi} e^{j \cdot 2 \cdot \pi \cdot f_D \cdot \cos(\theta) \cdot \tau} \, d\theta$$

$$R_g(\tau) = \sigma_g^2 \cdot J_0 \left(2 \cdot \pi \cdot f_D \cdot \tau\right) = \sigma_g^2 \cdot J_0(\beta \cdot v \cdot \tau) = \sigma_g^2 \cdot J_0 \left(2 \cdot \pi \cdot \frac{x}{\lambda}\right) \qquad (5.1.9)$$

where the second equality comes from the change of variable $v = f_D \cos(\theta)$ and the various equalities in the second line relate separation in time by τ and in space by x. By the way, we can also obtain the fading autocorrelation function by working directly in the space domain. See the Jakes complex gain generator in **Appendix B** for a derivation.

Sketch this autocorrelation function. Use normalized time or distance $u = f_D \tau = x/\lambda$.

$$u := -3, -2.98 .. 3$$

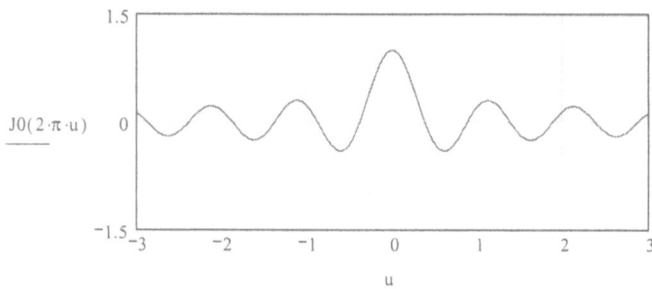

Normalized Time (Space) u=fDt=x/lambda

We can extract much useful information from the autocorrelation plot:

* The approximately periodic ripples reflect the quasiperiodicity we saw in the magnitude plots of gain vs time in the middle of **Section 3.1** and in **Appendix C**.

* The approximately periodic zeroes show that points separated in time (or space) by $f_D\tau$ (or x/λ) = 0.383, 0.878, ... are independent. That means that relatively closely spaced antennas can still provide diversity.

* The correlation dies away slowly because of the discontinuity and singularity in $S_g(f)$.

* The series expansion $J_0(\varepsilon)=1-\varepsilon^2/4+...$ gives a very useful approximation $R_g(u)=1-(\pi u)^2+...$ We'll use it later to show that Doppler-induced error floors in digital transmission vary as the square of $f_D T$.

Further to the last point, a more general series expansion for arbitrary power spectra is

$$R_g(\tau) \approx 1 - 2 \cdot \pi^2 \cdot v_{rms}^2 \cdot \tau^2 \qquad \text{(approx)} \qquad (5.1.10)$$

provided that the mean delay is zero. More about this in **Appendix F**.

Wide Sense Stationarity

If you have been keeping a skeptical eye on the derivation, you may have wondered why we could assume that $g(t)$ is wide sense stationary; that is, why

$$\frac{1}{2} \cdot E\left(\overline{g(t) \cdot g(t-\tau)}\right) = R_g(\tau) \qquad (5.1.11)$$

depends only on the time separation τ, not the absolute observation time t. For this, we return to **(3.3.2)** in Section 3.3, which implicitly defined $g(t)$ as

$$g(t) = \sum_i a_i \cdot e^{j \cdot 2 \cdot \pi \cdot f_D \cdot \cos(\theta_i) \cdot t} \qquad (5.1.12)$$

i

Substitution of this $g(t)$ into (5.1.11) gives

$$R_g = \frac{1}{2} \cdot E\left[\sum_i \sum_k a_i \cdot \overline{a_k} \cdot e^{j \cdot 2 \cdot \pi \cdot f_D \cdot \cos(\theta_i) \cdot t} \cdot e^{-j \cdot 2 \cdot \pi \cdot f_D \cdot \cos(\theta_k) \cdot (t-\tau)}\right]$$

$$= \frac{1}{2} \cdot E\left[\sum_i \sum_k a_i \cdot \overline{a_k} \cdot e^{j \cdot 2 \cdot \pi \cdot f_D \cdot (\cos(\theta_i) - \cos(\theta_k)) \cdot t} \cdot e^{j \cdot 2 \cdot \pi \cdot f_D \cdot \cos(\theta_k) \cdot \tau}\right] \tag{5.1.13}$$

If the scatterers at different Doppler shifts $f_D\cos(\theta)$ are uncorrelated over the $\{a_i\}$ ensemble, then we have

$$R_g(\tau) = \sum_i \sigma_i^2 \cdot e^{j \cdot 2 \cdot \pi \cdot f_D \cdot \cos(\theta_i) \cdot \tau} \tag{5.1.14}$$

which is a function only of the observation time difference τ. Thus $g(t)$ is wide sense stationary (WSS), a property in the observation time domain, if scatterers at different Dopplers are uncorrelated, a property in the Doppler spread domain. Note that we did not require that all scatterers be uncorrelated. In fact, the variance in (5.1.13) can be written

$$\sigma_i^2 = \frac{1}{2} \cdot E\left(\sum_{k'} a_i \cdot \overline{a_{k'}}\right) \tag{5.1.15}$$

where k' indexes all scatterers with Doppler shift $f_D\cos(\theta_i)$, which explicitly allows correlation among scatterers having the same Doppler shift.

Coherence Time

The time over which the complex gain stays roughly constant, so that the signal is relatively undistorted, is termed the *coherence time*. Clearly, the wider the Doppler bandwidth, the shorter the coherence time, since the time scales are reciprocally related, from Fourier theory. (In English: the wider the bandwidth, the faster the signal varies.) Generally, the fine structure dimension in one domain (time, here) is inversely proportional to the extent of the transform in the other domain (frequency, here). Because the assessments of degree of coherence and distortion are somewhat subjective and because the quantities depend on the specific functions, there is no universal definition. In fact, there is little point in trying to be really precise. However, one reasonable definition for coherence time is

$$T_c = \frac{1}{2 \cdot \pi \cdot v_{rms}} \tag{5.1.16}$$

Let's check it for the isotropic scattering case. From (5.1.8) and (5.1.15),

$$V_{rms} = \frac{f_D}{\sqrt{2}} \qquad T_c = \frac{1}{\sqrt{2} \cdot \pi \cdot f_D} \qquad\qquad (5.1.17)$$

Our plot of autocorrelation above normalized time with the Doppler frequency f_D, so we'll replot with scaled markers.

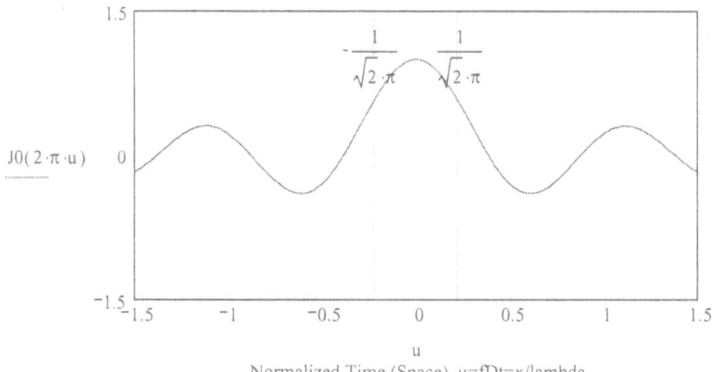

Normalized Time (Space) u=fDt=x/lambda

It's not a *bad* estimate. Of course, depending on what you want to do with the signals, it could be too large or too small.

5.2 Power Delay Profile and US Channels

Just as the **Doppler spectrum** represents the distribution of scatterers (density of power) in the Doppler domain ν, the power delay profile represents the distribution of scatterers (density of power) in the delay domain τ. It is central to any analysis or simulation of delay spread and intersymbol interference (ISI). In this section, we show its connection with the correlation between spaced carriers and link it to the uncorrelated scattering (US) assumption.

Spaced Frequency Correlation Function

To start, recall from **(3.3.3)** in Section 3.3 that the complex envelope output of a static channel is

$$y(t)=\sum_i g_i \cdot s(t-\tau_i) \qquad (5.2.1)$$

If the input is a sinusoid $\exp(j2\pi ft)$ at frequency f and we observe the output at a specific time $t=0$ we have the frequency response

$$y(0)=\sum_i g_i \cdot e^{-j\cdot 2\cdot\pi\cdot f\cdot\tau_i}=G(0,f) \qquad (5.2.2)$$

We want the *spaced-frequency correlation function*: the covariance between the responses due to sinusoids at f and $f-\Delta f$, when observed at the same time (say, $t=0$). Why? One reason is that it gives the distortion-free bandwidth, usually called the coherence bandwidth. For low-distortion transmission, we would like it to be quite broad in Δf. On the other hand, if we are trying to obtain diversity transmission, we want it to die away quickly in Δf.

The spaced-frequency correlation function (which is *not* necessarily real) is

$$C_g(\Delta f)=\frac{1}{2}\cdot E\left(G(0,f)\cdot\overline{G(0,f-\Delta f)}\right) \qquad (5.2.3)$$

Notation varies widely, so you may not see it as C_g in other works. It is shown as a function only of the spacing Δf, not of f itself. We'll see now that an assumption is required for this property to hold.

Uncorrelated Scattering

Substitution of (5.2.2) into (5.2.3) gives

$$C_g(\Delta f)=\frac{1}{2}\cdot E\left[\sum_i\sum_k \overline{g_i\cdot g_k}\cdot e^{-j\cdot 2\cdot\pi\cdot f\cdot\tau_i}\cdot e^{j\cdot 2\cdot\pi\cdot(f-\Delta f)\cdot\tau_k}\right]$$

$$\blacksquare = \frac{1}{2} \cdot E\left[\sum_i \sum_k g_i \cdot \overline{g_k} \cdot e^{-j \cdot 2 \cdot \pi \cdot f \cdot \left(\tau_i - \tau_k\right)} \cdot e^{-j \cdot 2 \cdot \pi \cdot \Delta f \cdot \tau_k}\right] \tag{5.2.4}$$

If the scatterers at different delays are uncorrelated - the uncorrelated scattering (US) assumption - then (5.2.4) reduces to

$$C_g(\Delta f) = \sum_k \left(\sigma_{g_k}\right)^2 \cdot e^{-j \cdot 2 \cdot \pi \cdot \Delta f \cdot \tau_k} \tag{5.2.5}$$

which is a function of Δf only, as we wanted. Thus the correlation between frequency responses depends only on the separation (a property in the input frequency domain f) if scatterers at different delays are uncorrelated (a property in the delay domain τ). Note that we did not require that all scatterers be uncorrelated. In fact the variances in (5.2.5) can be written

$$\left(\sigma_{g_i}\right)^2 = \frac{1}{2} \cdot E\left(\sum_{k'} g_i \cdot \overline{g_{k'}}\right) \tag{5.2.6}$$

where k' indexes all scatterers with delay τ_i, which explicitly allows correlation among scatterers having the same delay.

How realistic is the US assumption? Very. Recall that the complex amplitude g_k includes the phase shift due to the number of carrier wavelengths in the path length of scatterer k. That phase shift differs by many cycles if we compare paths at different delays τ_i and τ_k, so we can regard the phases as random, and the expectation $E\left(g_i \cdot \overline{g_k}\right) = 0$.

Power Delay Profile

If there are enough scatterers that we can use a density in t, then (5.2.5) becomes the Fourier pair

$$C_g(\Delta f) = \int_0^\infty P_g(\tau) \cdot e^{-j \cdot 2 \cdot \pi \cdot \Delta f \cdot \tau} d\tau \tag{5.2.7}$$

where the density $P_g(\tau)$ is termed the *power delay profile*. The total power in the scatterers is σ_g^2:

$$\sigma_g^2 = \int_0^\infty P_g(\tau) d\tau \tag{5.2.8}$$

For discrete scatterers, as in (5.2.5), we use impulses in the density:

$$P_g(\tau) = \sum_i \left(\sigma_{g_i}\right)^2 \cdot \delta\left(\tau - \tau_i\right)$$

(5.2.9)

Dominant scatterers, or clusters of scatterers, in the neighbourhood of the mobile show up as peaks in the delay power profile, as suggested by the sketches below. The first represents the same physical situation you saw in **Section 5.1**, where we examined the effect on Doppler spectrum. The second is the corresponding power delay profile, also annotated with numbers so you can relate the components in the two sketches. Note that "iso-delay contours" are ellipses with the mobile and the base as the two foci (see **Section 8.2**). You can see that the LOS component is the first to arrive. Also note that the delay of a component is unrelated to its Doppler shift, since, if it were displaced radially toward or away from the mobile, there would be no change in Doppler, but a substantial shift in delay.

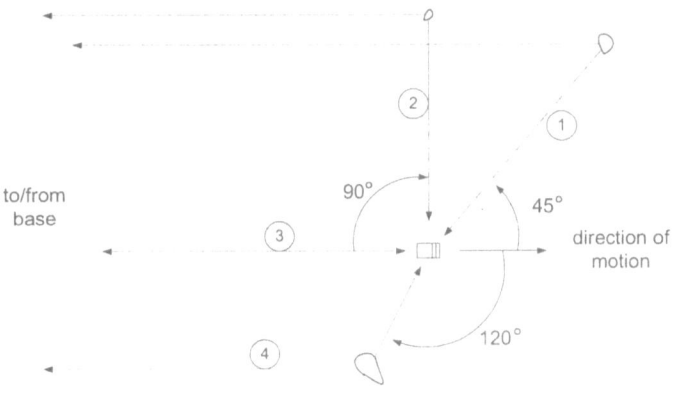

Example Power Delay Profiles

Some idealized power delay profiles are commonly used in analysis and design. Here we'll look at two examples: the exponential profile and the two-ray profile.

In urban settings, it is common for the power delay profile to be approximately exponential, with a corresponding spaced frequency correlation function having a first order Butterworth shape. Let's see it graphically.

⊡ Reference:D:\COURSES\MobChann\paperbook\Units.mcd(R)

$\sigma_g := 1$ $\tau_{rms} := 5 \cdot \mu s$ your choice of rms delay spread

power delay profile: frequency correlation function, from (5.2.7):

$$P_g(\tau) := \frac{\sigma_g^2}{\tau_{rms}} \cdot \exp\left(-\frac{\tau}{\tau_{rms}}\right) \qquad C_g(\Delta f) := \frac{\sigma_g^2}{\left(1 + j \cdot 2 \cdot \pi \cdot \Delta f \cdot \tau_{rms}\right)} \qquad (5.2.10)$$

$\tau := 0 \cdot \mu s, 0.1 \cdot \mu s .. 10 \cdot \mu s$ $\Delta f := -500 \cdot kHz, -495 \cdot kHz .. 500 \cdot kHz$

Here is the power delay profile defined in (5.2.10):

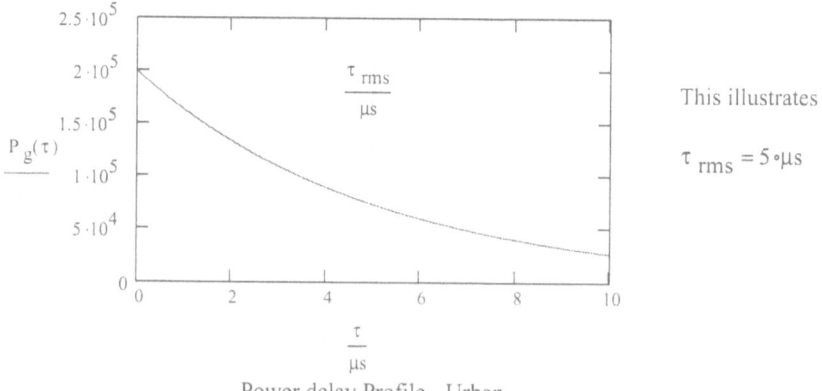

Power delay Profile - Urban

This illustrates

$\tau_{rms} = 5 \cdot \mu s$

and here is the corresponding frequency correlation function, also from (5.2.10)

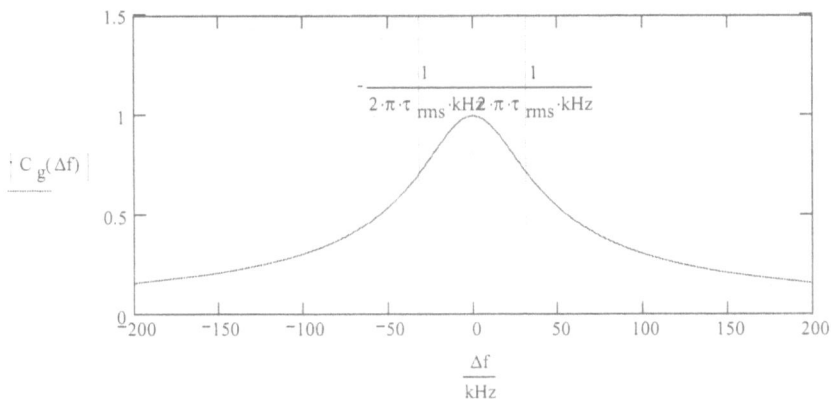

Spaced Frequency Correlation - Urban

You should see that for an rms delay spread of 5 μs - typical for downtown cores [**Sous94**] - carriers spaced by 100 kHz are almost completely decorrelated. Only very narrowband signals (about 10 kHz bandwidth or less) can get through relatively unscathed. However, for smaller delay spreads of, say, 1 μs, the distortion-free bandwidth is much larger.

A second power delay profile commonly used in analysis is the *two-ray model*. The arrivals are separated by the maximum delay spread τ_d and have variances $\sigma_{g0}{}^2$ and $\sigma_{g1}{}^2$. Your choice:

$$\tau_d := 2 \cdot \mu s \quad \text{also try } \tau_d = 1 \ \mu s$$

ratio of path powers $r = \sigma_{g1}{}^2 / \sigma_{g0}{}^2$ $\qquad r := 1 \qquad$ also try $r = 0.5$

$$\sigma_g := 1 \qquad \sigma_{g0} := \sqrt{\frac{1}{1+r} \cdot \sigma_g{}^2} \qquad \sigma_{g1} := \sqrt{\frac{r}{1+r} \cdot \sigma_g{}^2}$$

$$\tau_{rms} := \tau_d \cdot \frac{\sqrt{r}}{1+r} \qquad \text{rms delay spread (see (5.2.19) below)}$$

The power delay profile is a special case of the multiray profile (5.2.8)

$$P_g(\tau) = \sigma_{g0}{}^2 \cdot \delta(\tau) + \sigma_{g1}{}^2 \cdot \delta(\tau - \tau_d) \tag{5.2.11}$$

and for display purposes, we create

$$i := 0 .. 1 \qquad \tau_i := \tau_d \cdot i \qquad P_{g_0} := \sigma_{g0}{}^2 \qquad P_{g_1} := \sigma_{g1}{}^2$$

This is the two-ray power delay profile with your choice of parameters:

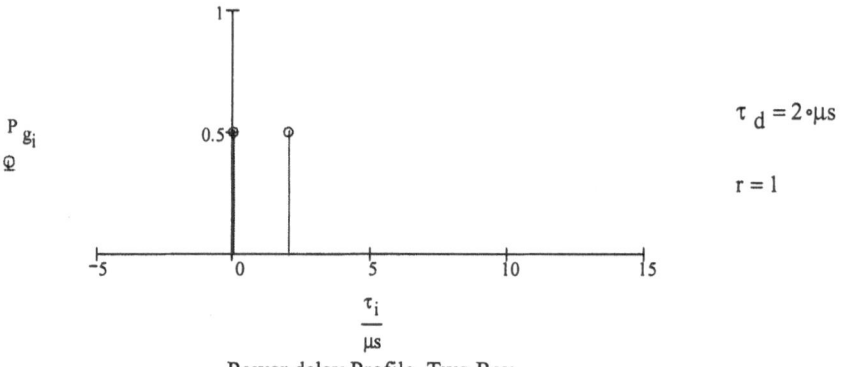

$$\tau_d = 2 \cdot \mu s$$

$$r = 1$$

Power delay Profile, Two Ray

The corresponding frequency correlation function is

$$C_g(\Delta f) := \sigma_{g0}^2 + \sigma_{g1}^2 \cdot e^{-j \cdot 2 \cdot \pi \cdot \Delta f \cdot \tau_d} \tag{5.2.12}$$

Spaced Frequency Corr'n, Two Ray

The spaced frequency correlation function is oscillatory, with fine structure dimension inversely proportional to the maximum delay spread τ_d, as we expect from Fourier theory.

Delay Moments

We have used the term *rms delay spread* several times in this discussion. We should define it and other properties of the power delay profile. First, the area:

$$\sigma_g^2 = \int_0^\infty P_g(\tau)\, d\tau \qquad (5.2.13)$$

which follows directly from the uncorrelated scattering assumption and (5.2.8). Next, the *mean delay*:

$$\tau_m = \frac{1}{\sigma_g^2} \cdot \int_0^\infty \tau \cdot P_g(\tau)\, d\tau \qquad (5.2.14)$$

Finally, the mean square delay spread is defined like a variance, hence rms delay spread like a standard deviation:

$$\tau_{rms}^2 = \frac{1}{\sigma_g^2} \cdot \int_0^\infty (\tau - \tau_m)^2 \cdot P_g(\tau)\, d\tau \qquad (5.2.15)$$

We had two specific power delay profiles above. For the exponential profile,

$$\tau_m = \tau_{rms} = \text{decay\_constant} \qquad (5.2.16)$$

For the two ray profile with maximum delay spread τ_d and

$$\sigma_g^2 = \left(\sigma_{g_0}\right)^2 + \left(\sigma_{g_1}\right)^2 \qquad (5.2.17)$$

We can obtain its moments in terms of τ_d and the ratio of average path powers

$$r = \frac{\left(\sigma_{g_0}\right)^2}{\left(\sigma_{g_1}\right)^2} \qquad (5.2.18)$$

This power delay profile is a special case of the multiray profile (5.2.8), from which we have the mean delay and rms delay spread as

$$\tau_m = \frac{r}{1+r} \cdot \tau_d \qquad\qquad \tau_{rms} = \frac{\tau_d}{1+r} \cdot \sqrt{r} \qquad (5.2.19)$$

As in the case of the fading autocorrelation $R_g(\Delta t)$ in **Section 5.1**, the second central moment gives an easy quadratic approximation if the time axis is shifted to give zero mean delay:

$$C_g(\Delta f) = 1 - 2 \cdot \pi^2 \cdot \tau_{rms}^2 \cdot \Delta f^2 \quad \text{(approx)} \qquad (5.2.20)$$

which is directly analogous to **(5.1.10)**. More about this in **Appendix F**.

Coherence Bandwidth

We used the term "distortion-free bandwidth" above. This idea is more formally termed the *coherence bandwidth*. It is related to the delay spread, since the fine structure dimension in one domain (frequency, here) is inversely proportional to the extent of the transform in the other domain (time, here). Because the assessments of degree of coherence and distortion are somewhat subjective and because the quantities depend on the specific functions, there is no universal definition. In fact, there is little point in trying to be really precise. However, one reasonable definition for coherence bandwidth is

$$W_c = \frac{1}{2 \cdot \pi \cdot \tau_{rms}}$$

(5.2.21)

or $2W_c$, if you are using RF bandwidths. The plots of spaced frequency correlation function above have markers at $\pm W_c$ and you can see that the coherence bandwidth, or fine structure dimension, is roughly W_c.

5.3 Scattering Function and WSSUS Channels

In the last two sections, we examined the second order statistics of the two phenomena separately:

* Time selective, flat fading channels (**Section 5.1**). We obtained the transform pair: gain autocorrelation function in the observation time domain t and Doppler spectrum in the Doppler shift domain ν. If scatterers at different Dopplers are uncorrelated, then the autocorrelation function depends only on the observation time difference; that is, the gain is wide sense stationary (WSS).

* Frequency selective, static channels (**Section 5.2**). We obtained the transform pair: spaced frequency correlation function in the input frequency domain f and power delay profile in the delay domain τ. If scatterers at different delays are uncorrelated - the uncorrelated scattering (US) assumption - then the spaced frequency correlation function depends only on the input frequency difference.

In both cases, we have a density function in one domain and an autocorrelation function in the other domain.

In this section, we'll tackle the second order statistics of frequency selective fading channels - those in which both phenomena are present. We'll see that the statistics are simplified if we invoke both the WSS and US assumptions to produce a double Fourier tranform pair: a joint density and a joint autocorrelation. Doing so introduces the important *scattering function*.

Time-Frequency Correlation Function

To begin, rederive the gain experienced by a carrier at input frequency f, observed at time t. For variety, we'll work with density functions, rather than discrete scatterers, in this analysis. From **(3.2.10)** of the discussion on the physical basis of fading in Section 3.2, we have the output as

$$y(t) = \int_{-\infty}^{\infty} \int_{-f_D}^{f_D} \gamma(\nu,\tau) \cdot e^{j \cdot 2 \cdot \pi \cdot \nu \cdot t} \cdot s(t - \tau) \, d\nu \, d\tau \tag{5.3.1}$$

where $\gamma(\nu,\tau)$ is the delay-Doppler spread function. For the particular case of single carrier input at frequency f, where $s(t)=\exp(j2\pi f t)$, this becomes

$$y(t) = e^{j \cdot 2 \cdot \pi \cdot f \cdot t} \cdot \int_{-\infty}^{\infty} \int_{-f_D}^{f_D} \gamma(\nu,\tau) \cdot e^{j \cdot 2 \cdot \pi \cdot \nu \cdot t} \cdot e^{-j \cdot 2 \cdot \pi \cdot f \cdot \tau} \, d\nu \, d\tau$$

$$\blacksquare = e^{j \cdot 2 \cdot \pi \cdot f \cdot t} \cdot G(t,f) \tag{5.3.2}$$

where the double Fourier transform in the second line comes from the summary **(I.15)** of the discussion on linear time variant filters in Appendix I. The gain we wanted is $G(t,f)$, the gain experienced by a carrier at input frequency f, observed at time t.

The covariance of present interest is between the gain experienced by a carrier at input frequency f, observed at time t, and the gain experienced by a carrier at input frequency $f-\Delta f$, observed at time $t-\Delta t$. It is in general a function of four variables: f, Δf, t and Δt. However, in anticipation of the simplifying WSSUS assumption, we'll write it as the *time-frequency correlation function*

$$R_G(\Delta t, \Delta f) = \frac{1}{2} \cdot E\left(\overline{G(t,f) \cdot G(t - \Delta t, f - \Delta f)} \right) \tag{5.3.3}$$

which depends only on the two variables Δt and Δf.

Calculating this one should be fun - we so rarely have a chance to write a quadruple integral. Substitute (5.3.2) into (5.3.3) and we have

$$R_G(\Delta t, \Delta f) = \int_{-\infty}^{\infty} \int_{-\infty}^{\infty} \int_{-\infty}^{\infty} \int_{-\infty}^{\infty} \text{integrand}\left(v_1, \tau_1, v_2, \tau_2 \right) dv_1 \, d\tau_1 \, dv_2 \, d\tau_2 \tag{5.3.4}$$

where

$$\text{integrand}\left(v_1, \tau_1, v_2, \tau_2 \right) = \blacksquare$$

$$\blacksquare = \frac{1}{2} \cdot E\left(\gamma(v_1, \tau_1) \cdot \overline{\gamma(v_2, \tau_2)} \right) \cdot e^{j \cdot 2\pi \cdot v_1 \cdot t} \cdot e^{-j \cdot 2\pi \cdot v_2 \cdot (t - \Delta t)} \cdot e^{-j \cdot 2\pi \cdot f \cdot \tau_1} \cdot e^{j \cdot 2\pi \cdot (f - \Delta f) \cdot \tau_2} \tag{5.3.5}$$

WSSUS Assumption

The situation in (5.3.5) looks desperate. However, recall that the WSS assumption (scatterers at different Dopplers are uncorrelated) and the US assumption (scatterers at different delays are uncorrelated) simplified the analysis in previous sections. Here we'll use *the WSSUS assumption* - that scatterers at different Dopplers *or* different delays are uncorrelated - and write the expectation in (5.3.5) as

$$\frac{1}{2} \cdot E\left(\gamma(v_1, \tau_1) \cdot \overline{\gamma(v_2, \tau_2)} \right) = S_\gamma(v_1, \tau_2) \cdot \delta(v_1 - v_2) \cdot \delta(\tau_1 - \tau_2) \tag{5.3.6}$$

where $S_\gamma(v, \tau)$ is called the *delay-Doppler power density function*, or just the *scattering function* This causes the quadruple integral to collapse to a relatively simple double Fourier transform

$$R_G(\Delta t, \Delta f) = \int_{-\infty}^{\infty} \int_{-\infty}^{\infty} S_\gamma(v, \tau) \cdot e^{j \cdot 2\pi \cdot v \cdot \Delta t} \cdot e^{-j \cdot 2\pi \cdot \Delta f \cdot \tau} \, dv \, d\tau \tag{5.3.7}$$

which depends only on the separations Δt and Δf.

The Scattering Function

The scattering function $S_\gamma(v,\tau)$ can be interpreted as the density of scattered power at Doppler shift v and delay τ, so that the received power in a differential element $dvd\tau$ at v, τ is $S_\gamma(v,\tau)\,dvd\tau$. Consequently, the scattering function is related to the delay power profile, the Doppler spectrum and the total power by

$$\int_{-\infty}^{\infty} S_\gamma(v,\tau)\,dv = P_g(\tau) \qquad \int_{-\infty}^{\infty} S_\gamma(v,\tau)\,d\tau = S_g(v) \qquad (5.3.8)$$

$$\int_{-\infty}^{\infty}\int_{-\infty}^{\infty} S_\gamma(v,\tau)\,dv\,d\tau = \sigma_g^2 \qquad (5.3.9)$$

Why have we bothered with all this math? What good does it do us? The answer is that you need the double autocorrelation function $R_G(\Delta t,\Delta f)$ to analyze the behaviour of any modulation, from OFDM to CDMA, over these difficult frequency selective fading channels. Eq. (5.3.7) lets you obtain that double autocorrelation function $R_G(\Delta t,\Delta f)$ from the double power density $S_\gamma(v,\tau)$. As for the double power density, or scattering function, it is linked in a straightforward way to the distribution of scatterers around the mobile: it is just the density of scatterers as a function of path delay and Doppler shift (i.e., the distance from the mobile and the azimuth with respect to the direction of motion, respectively).

What does the scattering function look like? Just as the scatterers affected **Doppler spectrum** and **power delay profile** in the previous sections, we can sketch the scatterering environment and the corresponding scattering function. For clarity, we use the same physical situation of isolated scatterers or clusters of scatterers as in the previous sections.

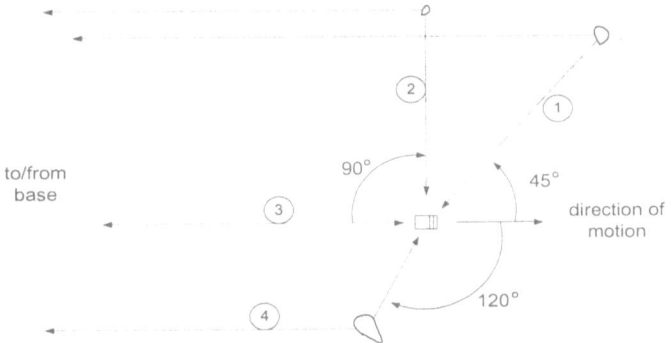

The corresponding scattering function is a surface giving the received power contributed from scatterers with delay τ and Doppler shift v. The sketch below is a top view of the peaks of that surface, so that you can relate it more easily to the physical environment sketched above.

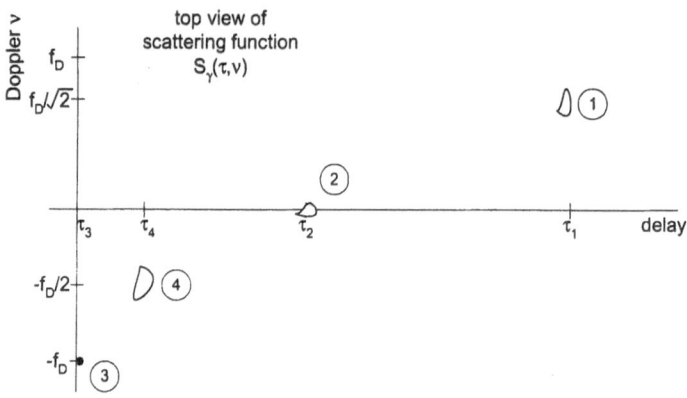

Note that location of peaks on the Doppler shift axis depends on the direction of motion. For example, if the mobile in the sketch above had been moving west, instead of east, then location of components along the vertical axis (the Doppler axis) would have been reversed. How would the scattering function look if the mobile had been moving north? What if it had been moving twice as fast?

Now let's see some typical measured scattering functions. You should be able to infer from these plots how the scatterers were distributed around the mobile.

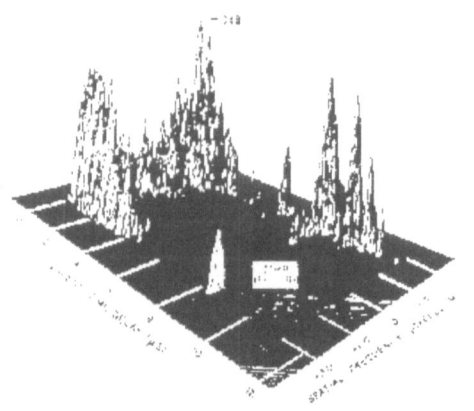

Click on this thumbnail to see a clearer picture.

Urban scattering function at 910 MHz
From [Cox73] (© 1973 IEEE)

And below is a suburban counterpart, but at a lower frequency:

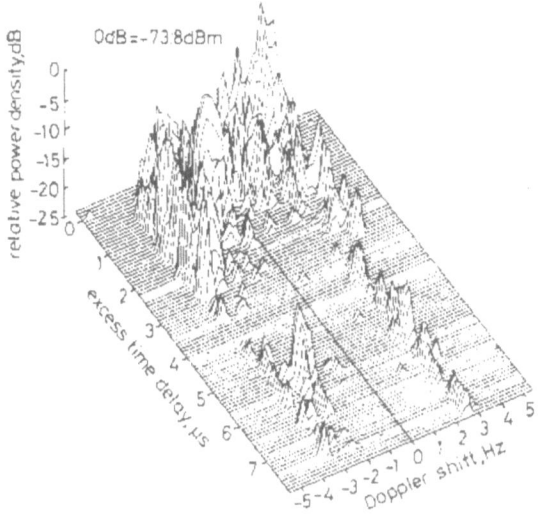

Suburban scattering function at 436 MHz with large delay spread.
From [**Pars82**], (© 1982 IEE, reprinted with permission)

Click on this
thumbnail to see a
clearer picture.

Separability

Here is a simplifying approximation. If the Doppler spectrum is not linked to the delay profile, then the scattering function is *separable*:

$$S_\gamma(\nu,\tau) = \frac{S_g(\nu)}{\sigma_g} \cdot \frac{P_g(\tau)}{\sigma_g} \tag{5.3.10}$$

so that we need only two one-dimensional functions to describe scattering, rather than one two-dimensional function. If we believe this one, we can immediately substitute it into (5.3.7), and obtain

$$R_G(\Delta t,\Delta f) = \frac{R_g(\Delta t)}{\sigma_g} \cdot \frac{C_g(\Delta f)}{\sigma_g} \tag{5.3.11}$$

which is a big help in analysis.

Separability may not be plausible in many cases. It is certainly not true of the sketched scattering function above (which was constructed that way to make that very point). Although the measured scattering functions above could reasonably be approximated as a product, there are many other examples of measured scattering functions that do not decompose neatly. Nevertheless, almost all analyses and simulations assume separability, because it is simpler and requires less channel information.

An Example

As a last step, we'll have a look at the scattering function, assuming separability, isotropic scattering and an exponential delay power profile - a *very* idealized model. However, it has some justification through interpretation as an ensemble average over many locations (see [**Sado98**]).

> ⊡ Reference:D:\COURSES\MobChann\paperbook\Units.mcd(R)

vehicle speed: $v := 100 \cdot \dfrac{km}{hr}$ $v = 27.778 \cdot \dfrac{m}{sec}$ $c := 3 \cdot 10^8 \cdot \dfrac{m}{sec}$

carrier frequency: $f_c := 900 \cdot MHz$ $\lambda := \dfrac{c}{f_c}$ $\lambda = 33.333 \cdot cm$

$$f_D := \frac{v}{\lambda} \qquad f_D = 83.333 \cdot Hz$$

delay spread: $\tau_{rms} := 5 \cdot \mu s$ $\sigma_g := 1$

From **(5.1.5)** and **(5.2.10)** the scattering function is the product

$$S_\gamma(v,\tau) := \sigma_g^2 \cdot \frac{1}{\pi \cdot f_D} \cdot \frac{1}{\sqrt{1 - \left(\dfrac{v}{f_D}\right)^2}} \cdot \frac{1}{\tau_{rms}} \cdot \exp\left(-\frac{\tau}{\tau_{rms}}\right) \qquad (5.3.12)$$

Generate points to a matrix and view the result:

$N_v := 50$ $i := 0 .. N_v$ $N_\tau := 20$ $k := 0 .. N_\tau$

$f_{Dmax} := 200 \cdot Hz$ $\tau_{rmsmax} := 10 \cdot \mu s$

$$v_i := \frac{i - 0.5 \cdot N_v}{N_v} \cdot 2.3 \cdot f_{Dmax} \qquad \tau_k := k \cdot 3 \cdot \frac{\tau_{rmsmax}}{N_\tau} \qquad S_{i,k} := S_\gamma(v_i, \tau_k)$$

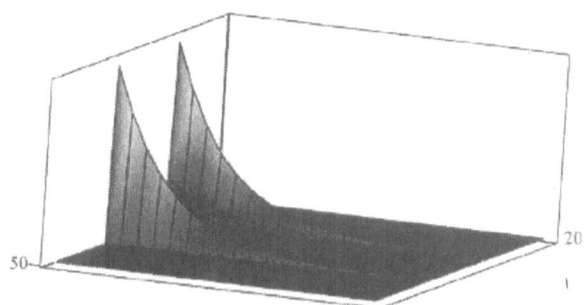

v left to right going into the page, τ left to right coming out of the page.

Have some fun. Use the mouse to drag the plot to other perspectives!

S

You can see a general similarity with the measured scattering functions displayed further above.

Here are some questions for you. What does the scattering function look like for the following more restrictive cases?

* flat fading (i.e., mobile in motion, narrowband modulation), with various selections of Doppler

* static, frequency selective channel (i.e., the ensemble of parked cars or of cellphones sitting on desks, all using wideband modulation), with various selections of delay spread

* static and flat (i.e., the ensemble of parked cars or of cellphones sitting on desks, all using narrowband modulation)

5.4 A TDL Channel Model for Analysis and Simulation

To this point, our channel descriptions have been based on signals defined on a continuous time axis, because it is obviously the case for RF and other analog signals. However, we frequently represent signals on a discrete time domain; for example:

* in simulations - the workhorses of communication system design;
* in DSP-based receivers (i.e., almost all receivers);
* in many analyses.

This section provides a model for the frequency-selective fading channel that we can use in discrete time situations. First, we'll see the model, because it it straightforward enough, then we'll review some of its limitations. Finally, we'll use it to generate some interesting sample plots showing the evolution in time of the frequency response.

The TDL Model

We consider the input signal $s(t)$ to be sampled at a rate f_s that is high enough to represent the output after Doppler spreading; that is, if the input signal has lowpass bandwidth W, then $f_s \geq 2(W+f_D)$. That way, we use the same sampling rate for input and output. The discrete time channel model is just a tapped delay line (TDL) with spacing t_s and time varying coefficients, as shown below:

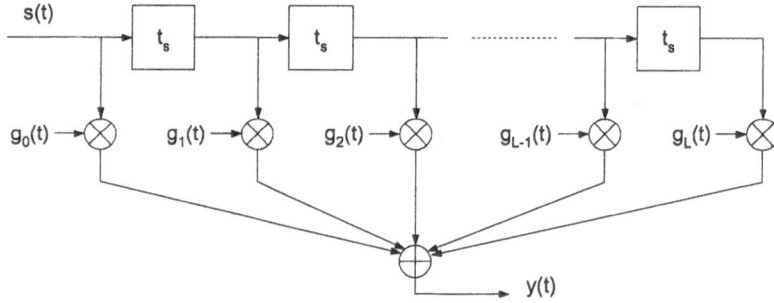

In keeping with the **WSSUS assumption**, the coefficients $g_i(t)$ are uncorrelated and, because of the central limit theorem, are Gaussian. The variances of the coefficients follow the **power delay profile**, so that

$$\left(\sigma_{g_i}\right)^2 = K \cdot \int_{i \cdot t_s}^{(i+1) \cdot t_s} P_g(\tau) \, d\tau \tag{5.4.1}$$

where K is a proportionality coefficient and the sampling interval is $t_s = f_s^{-1}$. The power spectra of the coefficients $g_i(t)$ do not have to have the same shape - that is, the **scattering function** does not have to be separable. However, separability is usually assumed, so that the power spectra have the same shape, scaled by the variances (5.4.1). As for the shape, it is usually (though not universally) taken to be the U-shaped spectrum of **isotropic scattering**. In summary, then, the coefficients $g_i(t)$ are mutually independent, and have autocorrelation functions

$$R_{gi}(\Delta t) = \left(\sigma_{g_i}\right)^2 \cdot J_0\left(2 \cdot \pi \cdot f_D \cdot \Delta t\right) \tag{5.4.2}$$

If there is a LOS component, it is placed as a DC, or mean value, on the first coefficient $g_0(t)$.

Limitations of the TDL Model

The model has its deficiencies. For one, as the signal bandwidth increases, the sampling rate must also increase, and the delay bin width t_s decreases. Fewer scatterers fall into each bin, and the Gaussian assumption for each coefficient $g_i(t)$ becomes increasingly suspect. Very wideband CDMA (15 to 20 MHz) can resolve down to individual paths, which we know to have roughly constant magnitude; further, very closely spaced paths at these bandwidths have been shown not to be independent [**Wu94**], though they may still be uncorrelated.

We can approach development of this model a little more formally through the sampling theorem. Because both input and output of the channel are bandlimited, they can be represented by their sample sequences, and reconstructed by

$$s(t) = \sum_k s(k \cdot t_s) \cdot p(t - k \cdot t_s) \qquad\qquad y(t) = \sum_i y(i \cdot t_s) \cdot p(t - i \cdot t_s) \tag{5.4.3}$$

where $p(t)$ is a lowpass interpolation pulse with a transform $P(f)$ that is flat for $|f| \le W + f_D$ and 0 for $|f| \ge f_s - W - f_D$. Candidates are Nyquist pulses like $sinc(f_s t)$ and raised cosine pulses, though the Nyquist property is not essential. The sampled channel output depends on the time variant impulse response through

$$y(i \cdot t_s) = \int_{-\infty}^{\infty} s(i \cdot t_s - \tau) \cdot g(i \cdot t_s, \tau) \, d\tau$$

$$= \sum_k s(k \cdot t_s) \cdot \int_{-\infty}^{\infty} p[(i - k) \cdot t_s - \tau] \cdot g(i \cdot t_s, \tau) \, d\tau$$

$$= \sum_j s[(i - j) \cdot t_s] \cdot \int_{-\infty}^{\infty} p(j \cdot t_s - \tau) \cdot g(i \cdot t_s, \tau) \, d\tau$$

$$= \sum_j s[(i - j) \cdot t_s] \cdot g_j(i \cdot t_s) \tag{5.4.4}$$

where $g_j(i t_s)$ in the last line is a notation for the integral in the line above. This form is just what we wanted - a discrete-time, time-varying model - and it corresponds directly to the TDL sketch above.

Unfortunately, the lowpassing implicit in the convolution of $p(t)$ and $g(t,\tau)$ means that we may not be able to resolve individual paths, which may in turn cause the various TDL coefficients to become correlated. To see this, calculate the covariance of coefficients j and m:

$$\frac{1}{2} \cdot E\left(g_j(i \cdot t_s) \cdot \overline{g_m(k \cdot t_s)}\right) = \blacksquare$$

$$\blacksquare = \frac{1}{2} \cdot \int_{-\infty}^{\infty} \int_{-\infty}^{\infty} p(j \cdot t_s - \tau) \cdot \overline{p(m \cdot t_s - \alpha)} \cdot E\left(g(i \cdot t_s, \tau) \cdot \overline{g(k \cdot t_s, \alpha)}\right) d\tau \, d\alpha$$

$$(5.4.5)$$

Assume WSSUS and separability to simplify it:

$$\frac{1}{2} \cdot E\left(g(i \cdot t_s, \tau) \cdot \overline{g(k \cdot t_s, \alpha)}\right) = P_g(\tau) \cdot R'_g\left[(i - k) \cdot t_s\right] \cdot \delta(\alpha - \tau) \qquad (5.4.6)$$

where the prime on $R'_g(it_s)$ indicates that it is normalized to unit variance. Substitute into (5.4.5), and we have the covariance as

$$\frac{1}{2} \cdot E\left(g_j(i \cdot t_s) \cdot \overline{g_m(k \cdot t_s)}\right) = R'_g\left[(i - k) \cdot t_s\right] \cdot \int_{-\infty}^{\infty} P_g(\tau) \cdot p(j \cdot t_s - \tau) \cdot \overline{p(m \cdot t_s - \tau)} \, d\tau \quad (5.4.7)$$

This shows that coefficients j and m are in fact correlated, and that the correlation increases with increasing delay spread. In other words, the independence of coefficients in the TDL model is expedient, but not completely accurate, even for a WSSUS channel.

An Example

We've spent enough space assessing the TDL model - now let's use it for something. As a first example, we'll create a transmitted signal, filter it and look at the resulting channel output. To avoid the effort of creating a typical signal (routines for doing that are in another of our minicourses), we'll just phase modulate a carrier with a cosine, to obtain a periodic, constant envelope signal:

    ➡ Reference:D:\COURSES\MobChann\paperbook\Units.mcd(R)

    ➡ Reference:D:\COURSES\MobChann\paperbook\Jakesgen.mcd(R)

Choose the modulation frequency, sampling rate and modulation index:

$$f_m := 10 \cdot kHz \qquad f_s := 32 \cdot f_m \qquad t_s := f_s^{-1} \qquad t_s = 3.125 \cdot \mu s \qquad f_d := 0.4$$

number of samples: $N_s := 100 \qquad i := 0 .. N_s - 1$

and here's the signal:

$$s_i := \exp\left(j \cdot 2 \cdot \pi \cdot f_d \cdot \cos\left(2 \cdot \pi \cdot f_m \cdot i \cdot t_s\right)\right) \tag{5.4.8}$$

Choose the Doppler frequency and the delay spread:

$$f_D := 1 \cdot Hz \qquad \tau_d := 10 \cdot \mu s$$

Assume a linear decay in the power delay profile:

$$P_g(\tau) := \frac{2}{\tau_d} \cdot \left(1 - \frac{\tau}{\tau_d}\right) \cdot (0 \cdot \mu s \le \tau) \cdot (\tau < \tau_d) \qquad \text{so the area } \sigma_g^2 = 1 \tag{5.4.9}$$

number of TDL taps in the model: standard dev'n per tap:

$$N_t := \text{ceil}\left(\frac{\tau_d}{t_s}\right) \qquad n := 0 .. N_t - 1 \qquad N_t = 4 \qquad \sigma_{g_n} := \sqrt{\int_{n \cdot t_s}^{(n+1) \cdot t_s} P_g(\tau) \, d\tau}$$

$$\tag{5.4.10}$$

$$\tau := -2 \cdot \mu s, -1.9 \cdot \mu s .. 1.5 \cdot \tau_d$$

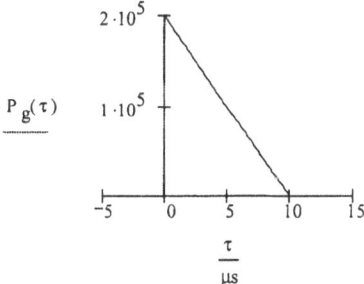

a linear delay power profile, unit area for convenience

Assume isotropic scattering, so use Jakes' model for generation of each TDL coefficient:

Initialize the parameter blocks of the N_t coefficients:

$$g^{<n>} := \text{Jakes\_init}(n) \tag{5.4.11}$$

At any observation time $i t_s$, coefficient n is given by:

$$\sigma_{g_n} \cdot \text{Jakes\_gen}\left(f_D \cdot t_s \cdot i, g^{<n>}\right) \tag{5.4.12}$$

so the channel output is the convolution

smaller$(N_t - 1, i)$

$$y_i := \sum_{n=0} s_{i-n} \cdot \sigma_{g_n} \cdot \text{Jakes\_gen}(f_D \cdot t_s \cdot i, g^{<n>})$$ (5.4.13)

Let's compare the input and output of the channel (reinitialize Jakes with F9 to see others):

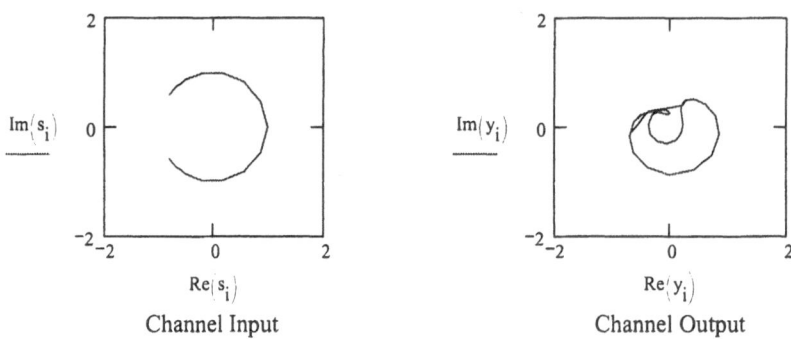

Channel Input   Channel Output

A delay spread of $\tau_d = 10$ µs causes considerable damage, even to a 10 kHz signal, but you can see that a Doppler of $f_D = 1$ Hz does not make the output aperiodic over this short observation interval. Try a much higher Doppler, and you'll see that the output drifts.

Another Example

Another interesting application is the evolution of the frequency response as the mobile moves through the field. We'll look at it across a fairly wide RF bandwidth

lowpass bw is half the RF bw:	tie signal sampling rate to bandwidth:		

$$W := \frac{300 \cdot \text{kHz}}{2} \qquad f_s := 8 \cdot W \qquad t_s := f_s^{-1} \qquad t_s = 0.833 \cdot \mu s$$

Choose the Doppler frequency and the delay spread:

$$f_D := 100 \cdot \text{Hz} \qquad \tau_d := 10 \cdot \mu s$$

Assume a linear decay in the power delay profile:

$$P_g(\tau) := \frac{2}{\tau_d} \cdot \left(1 - \frac{\tau}{\tau_d}\right) \cdot (0 \cdot \mu s \le \tau) \cdot (\tau < \tau_d) \qquad \text{so the area } \sigma_g^2 = 1$$

number of taps in the model:

$$N_t := \text{ceil}\left(\frac{\tau_d}{t_s}\right) \qquad n := 0..N_t - 1 \qquad N_t = 12$$

standard dev'n per tap:

$$\sigma_{g_n} := \sqrt{\int_{n \cdot t_s}^{(n+1) \cdot t_s} P_g(\tau) d\tau}$$

initialize the Jakes generator:

$$g^{<n>} := \text{Jakes\_init}(n)$$

and the frequency response at observation time it_s is

$$G(i,f) := \sum_n \sigma_{g_n} \cdot \text{Jakes\_gen}\left(f_D \cdot t_s \cdot i, g^{<n>}\right) \cdot e^{-j \cdot 2 \cdot \pi \cdot f \cdot t_s \cdot n} \qquad (5.4.14)$$

View with a frequency resolution of $1/8\tau_d$ and observation times separated by $N_{obs} := 500$ samples.

number of frequency samples:

$$N_f := \text{ceil}(2 \cdot W \cdot 8 \cdot \tau_d) \qquad N_f = 24$$

and the frequencies themselves:

$$k := 0..N_f \qquad f_k := -W + k \cdot \frac{1}{8 \cdot \tau_d}$$

$i := 0..15$ 15 snapshots, separated by N_{obs} samples or $N_{obs} \cdot t_s = 416.667 \cdot \mu s$

Calculate the time variant frequency response H(t,f):

$$H_{i,k} := G(i \cdot N_{obs}, f_k) \qquad H := \overrightarrow{|H|} \qquad \text{view the magnitude of H only}$$

This is the resulting time-variant frequency response. To see other examples, reinitialize the Jakes generator with F9 (or from the menu Math/Calculate Worksheet).

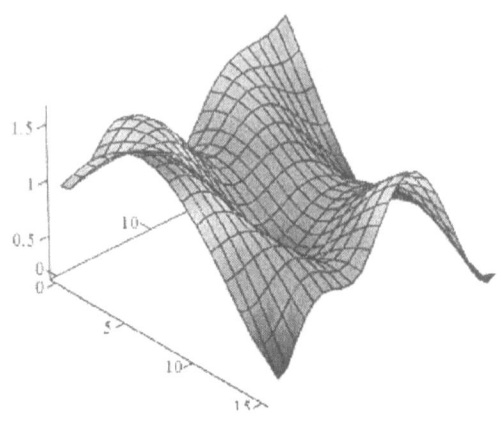

H

Drag the surface around
to different orientations
with your mouse.

Time runs left to right,
out of the page, from 0
to

$$15 \cdot N_{obs} \cdot t_s = 6.25 \cdot 10^3 \cdot \mu s$$

Frequency runs left to
right into the page, from

$$-\frac{W}{2} = -75 \cdot kHz \qquad \text{to}$$

$$\frac{W}{2} = 75 \cdot kHz$$

It's equally interesting as an **animation of the frequency and impulse responses**, so you can see nulls sweep through the band and the impulse response fluctuate. That's waiting for you in Appendix E.

6. CONNECTING FADING STATISTICS WITH PERFORMANCE

By now, you have absorbed a great deal of information on modeling channels for mobile and personal communications as randomly time-variant linear filters with Gaussian statistics. It's important not to lose sight of why you're going through all this - it's so that you can predict the performance of modulation and coding algorithms by analysis or simulation. Performance is the primary topic of two other texts in this series - *Detection and Diversity for Mobile Communications*, and *Channel Coding for Mobile Communications*. However, you will gain a preliminary exposure to performance analysis through the two topics in this section.

6.1 Random FM and Error Floor

At several points, these notes have mentioned an error floor, or irreducible error rate, caused by fading and the distortion it produces in the signal. With an error floor, you have transmission errors, even if there is no receiver noise at all. In this section, you will see how the rapid phase shifts in deep fades affect receivers that use a discriminator for detection, and how they cause problems for phase locked loops.

6.2 Level Crossing Rate and Error Bursts

You have already seen that transmission errors occur in bursts on the fading channel. Just how long are these bursts? How often do they occur? This section gives you some useful theory and two good rules of thumb for how the burstiness varies with changes in the average SNR.

6.1 Random FM and Error Floor

At several points, these notes have mentioned an error floor, or irreducible error rate, caused by fading and the signal distortion it produces. In this section, we'll focus on the random channel phase shifts, and calculate the error floor they produce in a receiver that uses a discriminator for signal detection. Discriminators respond to the time derivative of the phase of the received signal, one component of which is *random FM*, the derivative of the phase of the complex gain. We'll also look at the behaviour of phase locked loops in fading conditions. You can find discussion of random FM in [**Jake74, Lee82, Yaco93**].

Channel and Receiver Model

Suppose the transmitted signal has a constant envelope. Examples include FSK and PSK, which are still used today, as well as the newer CPM and GMSK modulations. At the receiver, we often use a discriminator, because it is inexpensive. If you don't remember what a discriminator is, review the text from your third or fourth year course in communications. Commonly used texts are [**Ziem76, Couc87, Taub71**].

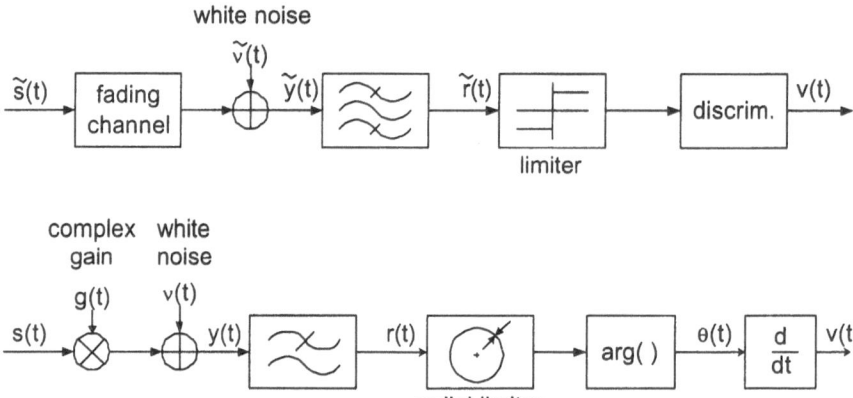

The sketches above show a typical link and discriminator receiver: first as an RF model in which a tilde indicates real bandpass signals; and then as a complex baseband model, where the corresponding complex envelopes are used. The transmitted signal is

$$s(t) = A \cdot e^{j \cdot \phi(t)} \tag{6.1.1}$$

It is distorted by flat fading and additive noise, so that the signal received from the antenna is

$$y(t) = g(t) \cdot s(t) + v(t) \qquad \text{where, in polar coordinates,} \qquad g(t) = a(t) \cdot e^{j \cdot \psi(t)} \tag{6.1.2}$$

The receiver first bandlimits the RF signal (i.e., lowpasses its complex envelope) to reduce the noise power. Its bandwidth is usually a compromise value: if it is too small, the filter distorts the signal; if it is too large, the filter admits too much noise. The result is

$$r(t) = g(t) \cdot s(t) + n(t) = m(t) \cdot e^{j \cdot \theta(t)} \tag{6.1.3}$$

After bandlimiting, the bandpass signal goes through a limiter, or infinite clipper. By removing the amplitude variation, it acts as an inexpensive version of automatic gain control. In complex gain terms, the output of the limiter (after implicit bandpass filtering to remove harmonics) is just $exp(j\theta(t))$. Finally, the discriminator produces the derivative of the phase, so that overall output is

$$v(t) = \frac{d}{dt}\theta(t) \tag{6.1.4}$$

In this discussion, we will be interested in the error floor, the residual error rate when the noise is zero. With zero noise, $\theta(t)=\phi(t)+\psi(t)$, and

$$v(t) = \phi'(t) + \psi'(t) \tag{6.1.5}$$

Since $\phi'(t)$ is the transmitted information, the complex gain phase derivative (known as random FM, for obvious reasons) appears as an additive disturbance.

Random FM Model

In the animation of complex gain in **Appendix D**, you saw that the channel phase can jump by almost 180 degrees during a fade, and it does so very quickly in deep fades. In effect, the mobile is hit twice: once by the loss of SNR, and once by the phase jump. We'll reproduce here a pair of graphs from that Appendix.

📩 Reference:D:\COURSES\MobChann\paperbook\Units.mcd(R)

📩 Reference:D:\COURSES\MobChann\paperbook\Jakesgen.mcd(R)

unit wavelength:	steps per wavelength	step size	sample index
$\lambda = 1$	$N_\lambda := 20$	$\Delta\lambda := \dfrac{\lambda}{N_\lambda}$	$k := 0 .. 11 \cdot N_\lambda$
$G := Jakes\_init(1)$	$g_k := Jakes\_gen(k \cdot \Delta\lambda, G)$		$x_k := k \cdot \Delta\lambda$
$g\,dB_k := dB\left[\left(\mid g_k \mid\right)^2\right]$	$i := 1 .. 11 \cdot N_\lambda$	$\Delta\psi_i := \arg\left(\overline{g_i \cdot g_{i-1}}\right)$	

The graphs of phase derivative and magnitude of the gain in dB (see next page) are particularly striking, since large peaks of derivative are almost always coincident with deep fades. The graph of phase derivative clearly shows that random FM is not Gaussian - it has a spiky character, rather like a sequence of impulses. Each spike corresponds to an approximate 180 degree phase reversal, as the complex gain speeds past the origin.

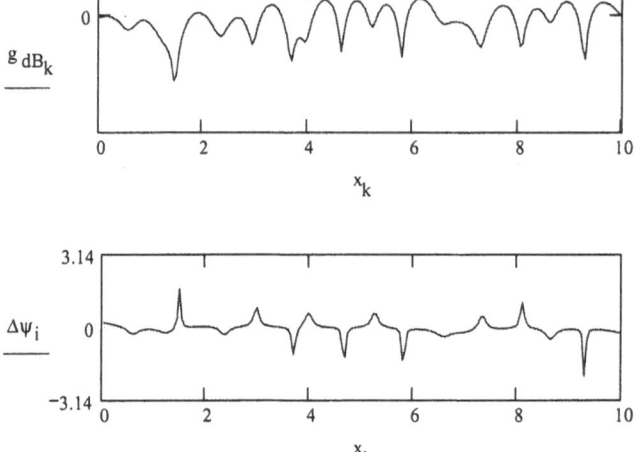

To see other
sample functions,
put the cursor on
the highlighted
initialization and
press F9.

The power spectrum of random FM is given in [**Jake74**] We have its pdf in Appendix G
as **(G.18)**, reproduced here as

$$P_{\psi'}(\psi') = \frac{1}{2} \cdot \frac{\sigma_g \cdot \sigma_{g'}^2}{\left[\sigma_{g'}^2 + \sigma_g^2 \cdot \psi'^2\right]^{1.5}} = \frac{1}{2} \cdot \frac{\left(2 \cdot \pi \cdot v_{rms}\right)^2}{\left[\left(2 \cdot \pi \cdot v_{rms}\right)^2 + \psi'^2\right]^{1.5}}$$

(6.1.6)

where we have also used **(F16)** from Appendix F. This is a very long-tailed distribution, as
you might expect from the graph. In fact, it has infinite variance!

FSK Example

We are running a mobile data link with binary FSK, frequencies $\pm f_d = 2400$ Hz, at 4800 bps.
The receiver detects the signal with a discriminator and no post-filter (i.e., no lowpass filter after
the discriminator). The system operates at 1.8 GHz, and vehicles travel at up to 100 km/h.
What is the irreducible bit error rate?

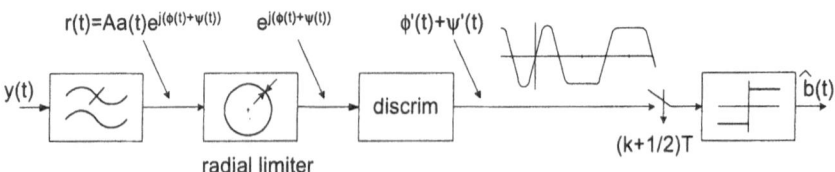

Start the analysis by noting that the discriminator output $v(t)$ would in principle be NRZ: rectangular pulses of duration 1/4800 sec and amplitudes $b2\pi f_d$, where $b=\pm 1$ is the data bit. In practice, smoothing at the transmitter to reduce the bandwidth and in the receiver's front end bandpass filter both act to round the pulse, so that it would reach these amplitudes only at the midpoint of the bit. Samples taken at the midpoint are therefore

$$v = b \cdot 2 \cdot \pi \cdot f_d + \psi' \tag{6.1.7}$$

since we are neglecting additive channel noise. If $b=-1$, then we make an error if $v>0$, or $\psi'>2\pi f_d$. From (6.1.6) the probability of error is

$$P_{floor} = \int_{2 \cdot \pi \cdot f_d}^{\infty} p_{\psi'}(\psi') d\psi' = \frac{1}{2} \cdot \left[1 - \frac{1}{\sqrt{1 + \left(\frac{v_{rms}}{f_d}\right)^2}} \right] \tag{6.1.8}$$

For isotropic scattering, **(5.1.8)** gives the rms bandwidth as $v_{rms}=f_D/\sqrt{2}$, so the error floor is

$$P_{floor}(f_D, f_d) := \frac{1}{2} \cdot \left[1 - \frac{1}{\sqrt{1 + 2 \cdot \left(\frac{f_D}{f_d}\right)^2}} \right] \tag{6.1.9}$$

We can see that the error varies as the square of the Doppler spread. This is quite characteristic of error floors in fading. In our example, we have

$$c := 3 \cdot 10^8 \cdot \frac{m}{sec} \qquad f_c := 1\ 8 \cdot GHz \qquad \lambda := \frac{c}{f_c} \qquad v := 100 \frac{km}{hr}$$

$$f_D := \frac{v}{\lambda} \qquad f_D = 166.667 \cdot Hz$$

$$f_d := 2400 \cdot Hz$$

Therefore the irreducible BER is

$$P_{floor}(f_D, f_d) = 2.394 \cdot 10^{-3}$$

Complete Error Rate Curve

This text does not put much emphasis on detection. However, you should know how the error floor shows up on typical curves of BER vs SNR. Essentially, there are two error-producing mechanisms, noise and random FM. For low SNR, the first one dominates, and we can ignore the floor; for high SNR, the second one dominates, and we can ignore noise.

In our example, the spacing between the tones makes them orthogonal. We already have an expression for the error rate of binary orthogonal signals, incoherently detected by matched filter in fading and noise, as in (4.3.5). On the grounds that this is very close to discriminator detection with an optimized prefilter, we will simply adopt it here. For a proper analysis, see [Korn90]. The error rate in the noise-limited region is therefore

$$P_{noise}(\Gamma_b) := \frac{1}{2 + \Gamma_b}$$ (6.1.10)

and the combined bit error rate is, roughly,

$$P_{comb}(\Gamma_b, f_D) := larger(P_{noise}(\Gamma_b), P_{floor}(f_D, f_d))$$ (6.1.11)

A graph illustrates the combined error rate for a selecton of Doppler spread values:

$$\Gamma_{bdB} := 10.. 50$$

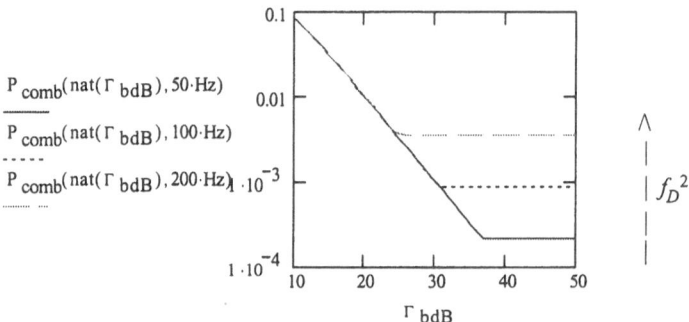

The BER curves that we will derive properly in the text *Detection and Diversity in Mobile Communications* in this series look much like this, except that the transition between regimes is rounded, rather than abrupt.

Effect on Phase Locked Loops

On static channels, PLLs are commonly used to extract carrier phase in order to allow coherent detection. (If you don't recall PLLs, look in almost any introductory text on communications, such as the ones cited at the beginning of this section). On fading channels, it is quite a different story. Random FM has a serious effect on PLL performance, as seen below.

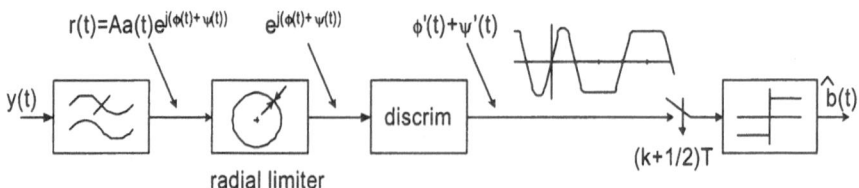

Consider transmission of unmodulated carrier, so the transmitted complex envelope $s(t)=1$. This could represent a pilot tone, for example. The PLL input is therefore

$$r(t) = g(t) + n(t) \qquad \text{with phase} \qquad \theta(t) = \psi(t) + \arg\left(1 + \frac{n(t)}{g(t)}\right) \qquad (6.1.12)$$

Recall the spiky appearance of the phase derivative. During the interval between spikes, while the derivative is small, we can assume that the PLL has pulled in. That is, the phase error allows the loop to operate near the stable zero crossing of its phase detector (PD) S-curve. Suddenly, a deep fade causes an abrupt reversal of phase. This puts the PLL at the metastable zero crossing of the phase detector characteristic, sometimes known as the ambiguity point. Since the error signal at the ambiguity point is very small, pulling in again is very slow. The result is a burst of errors produced by the loss of carrier phase reference, during and after the fade. Since these errors occur even without noise, we have an error floor. To make matters worse, the fade makes the SNR very low during the time the loop is trying to recover. This further slows the pull-in and increases the number of errors - and it can even cause a cycle slip if the loop pulls into the stable point of the next cycle. For these reasons, PLLs are rarely used for carrier recovery in mobile channels.

6.2 Level Crossing Rate and Error Bursts

General Considerations

At the end of **Section 4.3**, you saw that errors occur in bursts, since they are primarily confined to deep fades and seldom occur at other times. But how long are these bursts? How often do they occur? These are important issues for system design, where we must decide:

* Is a burst error correcting code useful?
* What type of burst error correcting code?
* What is the best length of an FEC (forward error correcting) block?
* What depth of interleaving is needed to break up bursts?
* In antenna switched diversity, how much delay is tolerable?

Unfortunately, precise results are not available - in part, because the definition of a burst is somewhat subjective (are you in a burst if the probability of error is high, but no actual errors have been made?). However, there are several useful results on the frequency and duration of fades, defined as intervals during which the channel gain is below a specified level. This section addresses that issue, and provides two useful rules of thumb for how the burstiness varies with changes in average SNR. Beyond that, and a few published analyses of finite interleaving depth, you are for the most part at the mercy of simulations.

It helps to have a picture. We'll look at a familiar graph below - fades in dB vs time.

for M samples spaced in time by $0.02/f_D$:

$$M := 400 \qquad i := 0..M \qquad fDt_i := 0.02 \cdot i \qquad \text{(normalized time)}$$

$$G := Jakes\_init(0) \qquad g_i := Jakes\_gen(fDt_i, G) \qquad MS\_value := 0 \qquad dB$$

You choose a fade level: $thresh := -6 \quad dB$

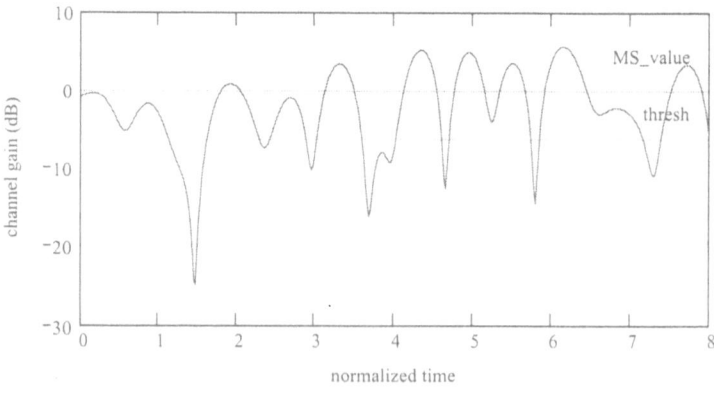

To see other complex gain patterns, recalculate the Jakes initialization with F9.

From the graph and other considerations, we can make a few observations:

* Deeper fades are both less frequent and shorter than shallower fades.

* We recall a rule of thumb from **Section 4.2** (following (4.2.6)) that the fraction of time spent 10 or more dB below the mean is 10%, 20 or more dB is 1%, etc. This governs the product of frequency and duration of the fades.

* The instantaneous system performance (BER, SNR_{out}, etc) may be a relatively soft function of instantaneous gain. However, it is useful to think in terms of thresholds, since some systems - coded systems and antenna switching, for example - behave that way. In effect, we identify poor performance with fades of a certain depth or more. This allows us to link system performance with level crossings of the channel gain.

Fade Rate (Level Crossing Rate)

This is badly presented in some texts. It's best to go back to the source [**Rice48**] for the underlying theory, a summary of which is presented below. What it means in fading channels is presented in [**Jake74,Lee82,Yaco93**].

The fade rate is a level crossing rate of the complex gain amplitude $r(t)=|g(t)|$. For a given amplitude threshold A, the number of upward-going (downward-going) crossings of $r(t)$ per second is

$$n_f(A) = \int_0^\infty r' \cdot p_{rr'}(A, r') \, dr' \tag{6.2.1}$$

where $r'(t)$ is the time derivative of $r(t)$. The proof, from Rice, can be obtained by examination of the sketch - the zoom view of a level crossing in the interval $(t, t+dt)$.

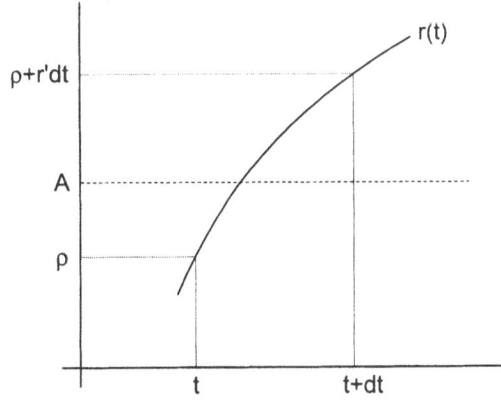

At times t and $t+dt$, $r(t)$ has values ρ and $\rho+r'dt$, respectively.

In this interval, there is an upward crossing if

$$A - \rho \leq r' \cdot dt \quad \text{and} \quad r' > 0 \tag{6.2.2}$$

Equivalently, we need

$$A - r' \cdot dt \leq \rho < A \tag{6.2.3}$$

The probability of an upward crossing in dt at t and a slope in dr' at r' is therefore

$$dr' \cdot \int_{A - r' \cdot dt}^{A} p_{rr'}(\rho, r') d\rho \tag{6.2.4}$$

so that the overall probability of an upward crossing is

$$P_{up} = \int_{0}^{\infty} \int_{A - r' \cdot dt}^{A} p_{rr'}(\rho, r') d\rho \, dr' = dt \cdot \int_{0}^{\infty} r' \cdot p_{rr'}(A, r') dr' \tag{6.2.5}$$

The second integrand is the expected number of crossings in any dt, that is, the crossing rate. Thus

$$n_f(A) = \int_{0}^{\infty} r' \cdot p_{rr'}(A, r') dr' \tag{6.2.6}$$

which concludes the proof.

We already have the joint pdf of r and r' as (G15) in **Appendix G**, reproduced here as

$$p_{rr'}(r, r') = \frac{r}{\sqrt{2 \cdot \pi \cdot \sigma_g^2 \cdot \sigma_{g'}}} \cdot \exp\left[-\frac{1}{2} \cdot \left(\frac{r^2}{\sigma_g^2} + \frac{r'^2}{\sigma_{g'}^2} \right) \right] \tag{6.2.7}$$

Substitution of (6.2.7) into (6.2.6) gives

$$n_f(A) = \frac{\sigma_{g'}}{\sigma_g} \cdot \frac{1}{\sqrt{\pi}} \cdot \frac{A}{\sqrt{2 \cdot \sigma_g^2}} \cdot \exp\left(\frac{-1}{2} \cdot \frac{A^2}{\sigma_g^2} \right) \tag{6.2.8}$$

Normalizing the threshold level by the rms level and substituting **(F16)** of Appendix F for $\sigma_{g'}$, we have

$$R = \frac{A}{\sqrt{2 \cdot \sigma_g^2}} \qquad \sigma_{g'} = 2 \cdot \pi \cdot \sigma_g \cdot v_{rms} \tag{6.2.10}$$

which gives

$$n_f(R) = 2 \cdot \sqrt{\pi} \cdot v_{rms} \cdot R \cdot e^{-R^2} \qquad \text{fades per second} \tag{6.2.11}$$

$$\blacksquare = \sqrt{2 \cdot \pi} \cdot f_D \cdot R \cdot e^{-R^2} \qquad \text{for isotropic scattering, \textbf{\underline{(5.1.8)}} of Section 5.1}$$

We see from (6.2.11) that the level crossing rate is proportional to the rms Doppler spread. It also has a very interesting asymptotic form for low thresholds, as illustrated in the graph below.

$$\text{RdB} := -40 .. 10 \qquad \text{normrate}(R) := 2 \cdot \sqrt{\pi} \cdot R \cdot e^{-R^2} \qquad \text{(normalized by rms Doppler)}$$

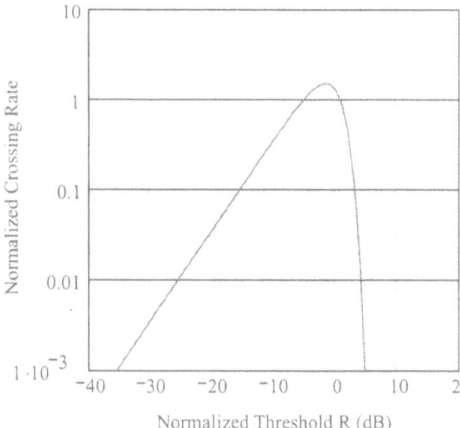

Points of interest:

* Maximum crossing rate is at a level 3 dB below the mean ($R^2 = 0.5$).

* The fade rate is proportional to f_D.

* For about -5 dB and lower, a 6 dB decrease in level (R cut in half) means half the fade rate.

Fade Duration

The average fade duration is easy, now that we have the other relations. Denote it by $t_f(R)$. Then

$$\text{Prob}\left(\frac{r}{\sqrt{2 \cdot \sigma_g^2}} \leq R \right) = t_f(R) \cdot n_f(R) \tag{6.2.12}$$

From **(4.2.6)** of Section 4.2, we have the Rayleigh cdf as $1 - exp(-R^2)$. Therefore the average fade lasts, in seconds,

$$t_f(R) = \frac{1 - e^{-R^2}}{n_f(R)} = \frac{1 - e^{-R^2}}{2 \cdot \sqrt{\pi} \cdot v_{rms} \cdot R \cdot e^{-R^2}} = \frac{1 - e^{-R^2}}{\sqrt{2 \cdot \pi} \cdot f_D \cdot R \cdot e^{-R^2}} \qquad \begin{array}{l}\text{(the latter for isotropic} \\ \text{scattering)}\end{array} \tag{6.2.13}$$

Notice that for small R, the asymptotic form is just

$$t_f(R) = \frac{R}{\sqrt{2 \cdot \pi \cdot f_D}} \quad \text{seconds}$$ (6.2.14)

The average fade duration is inversely proportional to rms Doppler spread. Normalizing it should give an interesting plot:

$$\text{normdur}(R) := \frac{1 - e^{-R^2}}{\sqrt{2 \cdot \pi \cdot R \cdot e^{-R^2}}} \qquad \text{nat}(x) := 10^{0.1 \cdot x}$$ (6.2.15)

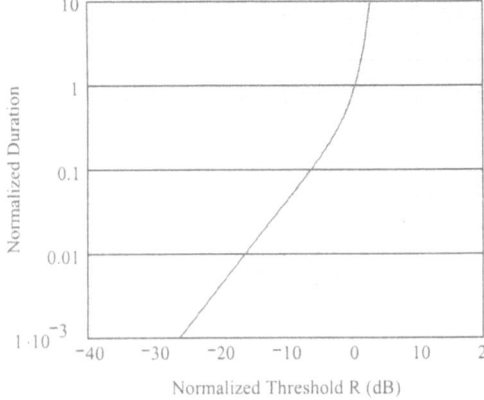

Points of interest:

* Fade duration is proportional to amplitude threshold level for -5 dB or lower.

* For about -5 dB and lower, a 6 dB decrease in level (R cut in half) mean fades are half as long.

* The fade duration is inversely proportional to f_D.

More About the Fades

From the above, we have the handy rules of thumb:

  * If the fade depth threshold is reduced by 6 dB (a factor of 4), then the fades are half as long *and* half as frequent. Extension to other reductions is obvious.
  * Increasing the transmitter power by 6 dB results in half as many error bursts, averaging half as long.

We have seen the averages for fade rate and fade duration. Is there any information about their pdfs? Not much. But here are some useful results:

* The intervals between fades (intervals over threshold) are approximately exponentially distributed for $R<-15$ dB or so [**Arno83**]. Since this level also implies short fades, we can say that the occurrence of deep fades has a Poisson distribution.

* The distribution of fade duration is not exponential. For deep fades [**Bold82**]:

$$\text{Prob}(\text{fade\_length} > u \cdot t_f) = \frac{2}{u} \cdot I_1\left(\frac{2}{\pi \cdot u^2}\right) \cdot \exp\left(-\frac{2}{\pi \cdot u^2}\right)$$

where u is the fade length normalized by the mean length t_f. It drops approximately as u^{-3} for $u>1$ (fades longer than the mean fade duration)

Try It

Now go back to the graph of fading magnitude at the top of this section and try various threshold levels. See if the fade rates and durations agree with the rules of thumb we just derived. Try many sample functions, using F9.

Simple Example

In a mobile data system, what does a 3 dB increase in transmitter power accomplish in terms of BER and error burst statistics?

* The BER is cut in half, since BER depends inversely on SNR (**Section 4.3**).

* Bursts are 71% as long and 71% as frequent (i.e., 1/sqrt(2) for each).

A More Interesting Example

A mobile data system operates at 1 GHz, with vehicles traveling at 108 km/h. The data rate is 9600 bps using FSK, detected by discriminator. The average error rate is 10^{-3}. Estimate (roughly) the burst duration and frequency, both in bits.

* Assume the BER is that of binary orthogonal signals, incoherently detected, so (**Section 4.3** again)

$$P_{e\_av} = \frac{1}{2 + \Gamma_b}$$

Since our error rate is 10^{-3}, the average SNR per bit is

$$\Gamma_b = 30 \cdot dB = 10^3$$

* Make a major approximation: bursts are defined by instantaneous BER≥ 0.1. This defines the threshold value of instantaneous SNR γ_b as

$$\frac{1}{2} \cdot \exp\left(-\frac{\gamma_b}{2}\right) = 0.1 \qquad \text{from which} \quad \gamma_b = 3.22 \quad \text{or 5 dB}$$

This corresponds to a 25 dB fade, so that, from the results above

$$\frac{n_f}{f_D} = 0.14 \qquad\qquad t_f f_D = 0.022$$

* Since $f_D = v/\lambda = 100$ Hz, we have

$$n_f \approx 14 \cdot Hz \quad \text{deep fades/sec, averaging } t_f \approx 224 \cdot \mu s \quad \text{each}$$

or, in terms of bits,

$$\text{interfade interval} = \frac{9600}{14} = 686 \quad \text{bits}$$

$$\text{fade duration} = 9600 \cdot 224 \cdot 10^{-6} = 2.2 \quad \text{bits}$$

This is like random (Bernouilli) errors, except for little clusters.

Last Example

This is the same as the example above, but the vehicle is moving at 50 km/h. We could follow the same procedure, or just scale the previous result by the ratio of Dopplers (equivalently, the ratio of vehicle speeds). Therefore

$$\text{average interfade interval} = 686 \cdot \frac{108}{50} = 1482 \quad \text{bits}$$

$$\text{average fade duration} \approx 2.2 \cdot \frac{108}{50} = 5 \quad \text{bits}$$

7. A GALLERY OF CHANNELS

Designs of modulation, coding and entire systems are strongly affected by the propagation environment. Optimization for urban conditions, for example, does not imply good performance in rural or indoor settings. To this point in the course, we have been concerned with quite general models for propagation, ones which exposed fundamental properties and time-frequency relationships. Now it is time to examine the behaviour of a few specific and commonly used channels. In this section we look briefly at three of them:
* macrocells;
* urban microcells;
* indoor microcells.

It is not the objective of this text to provide a comprehensive summary of the many reported propagation studies. Instead, you will see a few highlights. For more detail, consult the cited references.

In characterizing the channels, our primary interest lies in:

* **power delay profile**;
* **Rice K factor**;
* **path loss exponent**;

7.1 Macrocells

Macrocells are used in first- and second-generation cellular systems, such as AMPS, D-AMPS (also known as NADC and IS-54) and GSM. A typical cell radius is about 10 km, although cells can be significantly smaller in downtown cores.

7.2 Urban microcells

For the very high traffic volume expected of personal communications, reuse of the time-frequency resource will have to be intensive, forcing much smaller cells. These microcells range in radius from a few hundred metres to 1 km.

7.3 Indoor picocells

Within an office or large building environment, the reflections and absorptions of RF power by walls and floors, the strongly rectangular orientation of the obstacles and the slow movement of mobiles and scatterers (read people) make quite a different propagation environment from the two outdoor channels seen above.

7.1 Macrocells

Macrocells are used in first- and second-generation cellular systems, such as AMPS, D-AMPS (also known as NADC and IS-54) and GSM. A typical cell radius is about 10 km, although cells can be significantly smaller in downtown cores. The base station antenna is usually located on a vantage point - a tower or tall building - about 30 m to 100 m high, so that it has an unimpeded 360 degree view.

Impulse and Frequency Responses

Below are a typical urban impulse response magnitude $|g(t_o,\tau)|$ and corresponding frequency response magnitude $|G(t_o,f)|$. They are"snapshots", since they are at a particular observation time t_o.

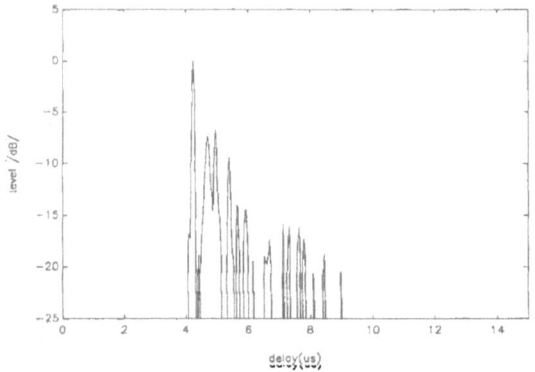

Click on this thumbnail to see a clearer picture.

Typical urban impulse response magnitude.
From [**Sous94**] (© 1994 IEEE)

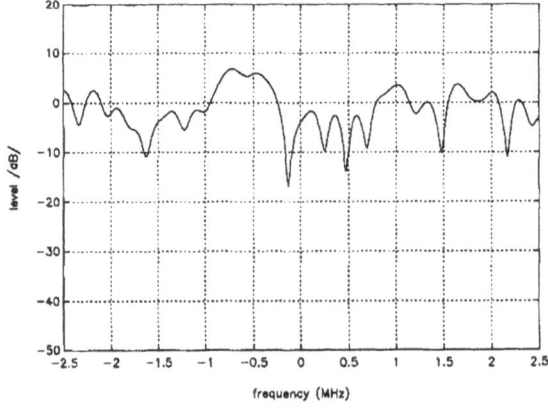

Click on this thumbnail to see a clearer picture.

Corresponding frequency response magnitude.
From [**Sous94**] (© 1994 IEEE)

Power Delay Profile

Measurements in urban areas [**Sous94**] show a roughly exponential power delay profile. Somewhat better accuracy in modelling is achieved with a few clusters, each with a delay and an exponential power delay profile [**Hash79**]. The snapshot impulse response magnitude above could have been drawn from either model. Other measurements in a suburban area at 436 MHz [**Bajw82**] produced the following power delay profile $P_g(\tau)$ and corresponding spaced-frequency correlation function $C_g(\Delta f)$ (see **(5.2.7)**).

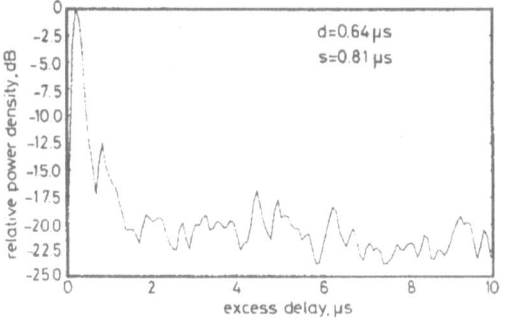

Click on this thumbnail to see a clearer picture.

Typical suburban power delay profile.
From [**Bajw82**] (© 1982 IEE, reprinted with permission).

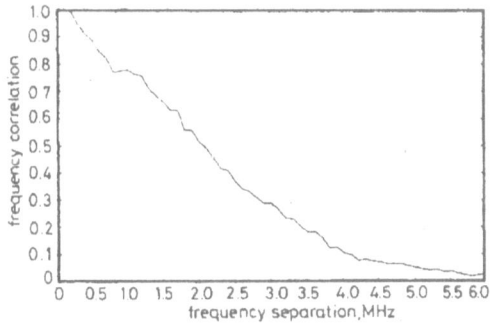

Click on this thumbnail to see a clearer picture.

Corresponding frequency correlation function.
From [**Bajw82**] (© 1982 IEE, reprinted with permission).

The rms delay spread **(5.2.15)** is a very approximate summary of the full power delay profile. However, it is an adequate characterization in Rayleigh fading for narrowband systems if τ_{rms} is less than about 0.2 symbol times [**Deng95**, **Cave95**] since the signal bandwidth is then not sufficient to resolve details of the profile. This simplification may not hold for Rice fading, however.

The rms delay spread varies by location, and its complementary cumulative distribution function in an urban area is shown below.

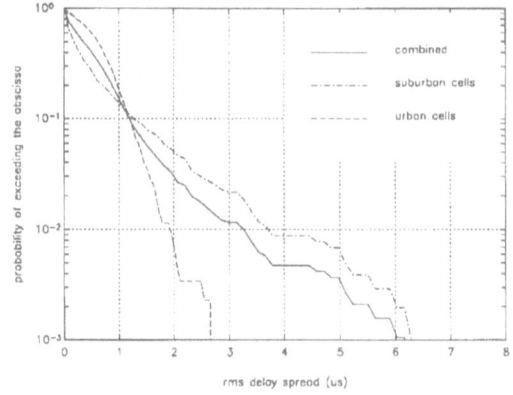

Click on this thumbnail to see a clearer picture.

Complementary cumulative distribution function of rms delay spread.
From [**Sous94**] (© 1994 IEEE).

Delay spread is affected by directionality of the transmit and receive antennas (see **Section 8**). The plot below shows that even 120 degree sectorization of a base station antenna reduces delay spread to about 2/3 the value experienced with an onmidirectional antenna.

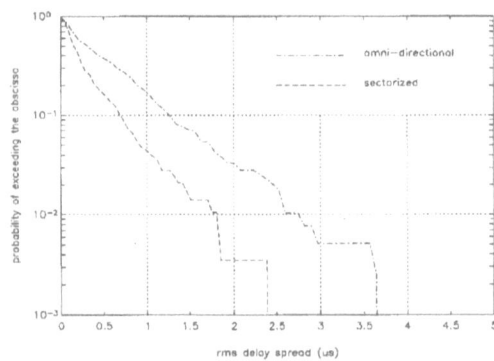

Click on this thumbnail to see a clearer picture.

Effect of base station antenna directionality on rms delay spread.
From [**Sous94**] (© 1994 IEEE).

One variation on the macrocell theme is that of fixed wireless access, in which the "mobiles" are fixed terminals, usually in homes or businesses, to be provided with wideband wireless access to a backbone, such as the Internet. Here the terminals can be equipped with directional antennas and, as you will see in **Section 8.2**, directionality has a significant effect on delay spread. One major measurement project [**Erce99**] showed that the resulting power delay profile is well modeled by a "spike plus exponential" shape; that is, a strong, discrete first arrival, followed by scattered power at a lower level that declines exponentially with delay.

Rice K Factor

An interpretation by [**Dari92**] of two propagation studies gives the following range of K factors:

* urban centre - high building density (>30%): $K=-\infty$ dB, i.e., Rayleigh fading;
* urban area - moderate building density (20%-30%): $K=0$ dB to 4 dB;
* urban - low building density (10%-20%): $K=4$ dB to 6 dB;
* suburban: $K=6$ dB to 10 dB
* open, rural: $K>10$ dB.

The amplitudes of individual paths are reported in [**Hash79**] to be lognormal over large areas (i.e., the shadowing variations). Locally, the initial path amplitudes are **Nakagami** (a generalization of the Rayleigh and Rice pdfs), and the following paths are lognormal in amplitude. The composite set of paths was shown in [**Bult89b**] to follow closely a Rayleigh distribution, as reproduced below:

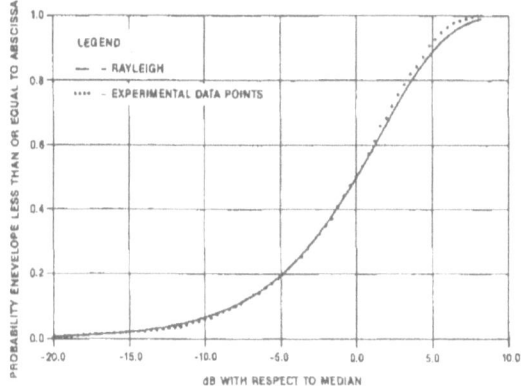

Click on this thumbnail to see a clearer picture.

Cumulative distribution function computed from a relatively small macrocell. From [**Bult89b**] (© 1989 IEEE).

On the other hand, measurements on a fixed wireless link in a typical urban macrocell [**Mich99**] show that the K factor in successive time segments of 15 min is itself Gaussian distributed with a mean of about 8 dB.

Path Loss Exponent

For typical urban settings, [**Hata80**] summarized measured path loss as having an exponent in the range 3.6 to 4. This is, of course, a very approximate way to characterize propagation, since so much depends on the details of the physical environment that we lump into the shadowing phenomenon.

An excellent summary of several path loss mechanisms and models can be found in [**Yaco93**]. Discussion of various measurement studies and their limits of applicability is contained in [**Stee92**].

7.2 Urban microcells

For the very high traffic volume expected of personal communications, reuse of the time-frequency resource will have to be intensive, forcing much smaller cells. These microcells range in radius from a few hundred metres to 1 km. The base antennas are correspondingly lower, mounted on the sides of buildings or on lampposts.

Path Loss and K Factor

As noted in **Section 1** of these notes, a simple two-ray model demonstrates the presence of two regimes: near the base station, where the path loss exponent is relatively small (on the order of 2), and the signal level fluctuates significantly from the interference between received components; and far from the base station, where the exponent is much greater (theoretically, 4), because of the combination of inverse square law loss and persistent destructive interference between the components. The break point, measured in wavelengths, is between 4 and 8 times $h_b h_m / \lambda^2$, the product of base and mobile antenna heights, each normalized by wavelength. If the mobile, or handheld unit, has an antenna 5λ above the ground, then a 5 meter high base antenna puts the break point at about 150 m. This dual behaviour is ideal for a microcell with radius near the break point: the *mean* power drops slowly to the cell edge (although any specific cell has strong fluctuations superimposed), and falls off rapidly beyond that point, thereby minimizing interference in cochannel cells.

The measurements below, from [**Taga99**], confirm the presence of two regimes in the true, and much more complex, environment. They show a distinct break point, in which the regression lines are very close to the inverse square and inverse fourth power laws.

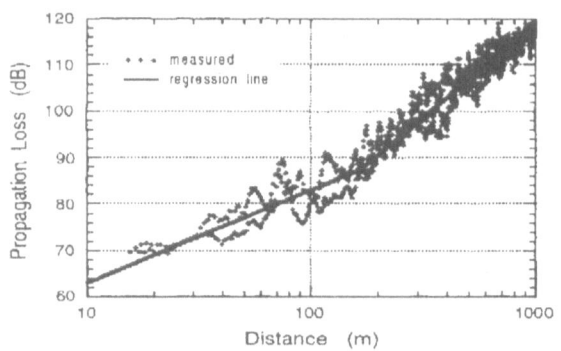

Click on this thumbnail to see a clearer picture.

Propagation loss characteristics of a typical urban road.
From [**Taga99**] (© 1999 IEEE).

In a useful paper, [**Gree90**] reports on extensive urban microcellular measurements at 905 MHz and demonstrates a close fit to a double regime. The inner exponent was normally close to 2, and the outer exponent ranged from 4.4 to 9.2, depending strongly on the the particular environment. The break point was typically 100 to 200 m, which is reasonably close to the **two-ray breakpoint value** of 250 m for the base and mobile antenna heights of 7.5 m and 1.5 m, respectively. [**Stee92**] cites and interprets measurement studies by [**Gree90**] and [**Chia90**], among others, that support the two regime model.

Rice K Factor

Also cited in [**Stee92**] are the [**Gree90**] measurements of microcell *K* factor, reproduced below. Although these values of *K* seem very favourable, the discussion in [**Stee92**] concludes that the *K* factor is too erratic for statistical modeling, and that designers should work to worst case conditions.

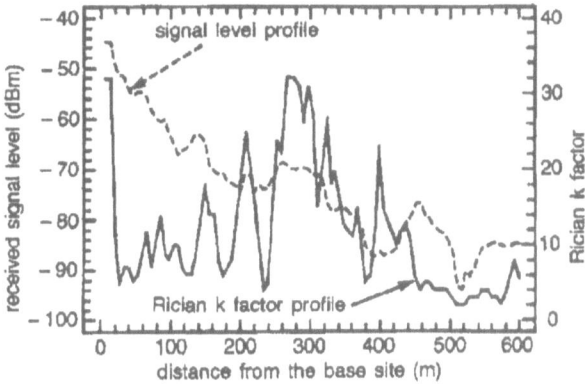

Click on this thumbnail to see a clearer picture.

Signal level and *K*-factor in an urban microcell. From [**Gree90**] (© 1990 *BT Technology Journal*, reproduced with permission).

In another measurement and modeling study of urban microcells, [**Bult89b**] demonstrates a close fit to the Rice distribution, as shown below. The illustration below shows that *K*= 7 dB for one city block. However, another nearby block had *K*= 12 dB, which reinforces Steele's cautionary comment.

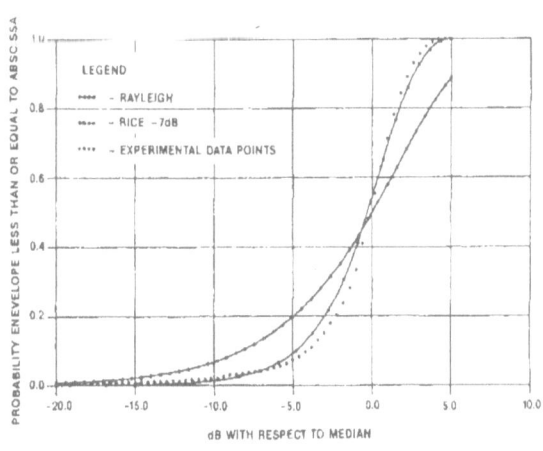

Click on this thumbnail to see a clearer picture.

Cumulative distribution functions are Rician for a microcell and Rayleigh for a small macrocell.. From [**Bult89b**] (© 1989 IEEE).

Power Delay Profile

The **power delay profile and rms delay spread** were extracted from measurements by [**Bult89b**]. This study showed a roughly exponential power delay profile, with an rms delay averaging 0.48 µs. A study of LOS urban microcells [**Taga99**] also showed small delay spreads from about 100 ns to about 400 ns, with a non-monotonic dependence on distance from the base, as illustrated below:

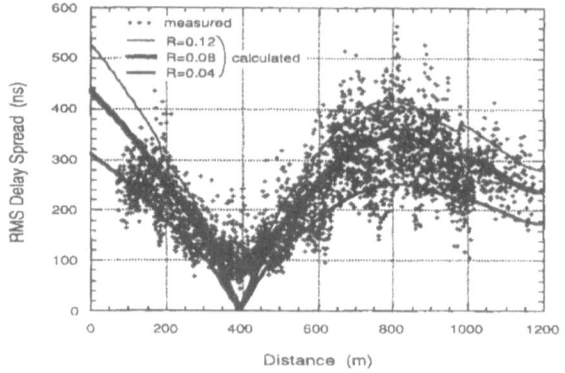

Click on this thumbnail to see a clearer picture.

Comparison of calculated and measured delay spread in Chuo Avenue, Tokyo. From [**Taga99**] (© 1999 IEEE).

We see that microcells are a more benign environment than **macrocells** for symbol-by-symbol signal detectors, both in their delay spread and their path loss exponents. The wider signal bandwidths of microcellular PCS systems may nevertheless span one or more coherence bandwidths, forcing the use of equalizers or MLSE receivers. There is some benefit in this; the more sophisticated receivers can exploit the delay spread to gain a form of diversity, as we will see in the text *Detection and Diversity* in this series.

CDMA as a wideband modulation for PCS in microcellular systems also makes explicit use of diversity in the delay spread. Since such systems allow any form of SNR improvement to be exploited in a uniform fashion, and are usually designed to operated close to the edge of capacity, it is important to understand the limits of diversity and the WSSUS assumption. [**Wu94**] reports that for CDMA systems that resolve the paths, 80% of the received power lies in the strongest path at 5 MHz bandwidth, in the 2 strongest at 10 MHz and the 3 strongest at 20 MHz. At 5 MHz, the first 3 paths are uncorrelated, but the second two are relatively weak; at 20 MHz, the first 3 paths are closer to the same power, but are correlated, so there is less diversity to be gained.

7.3 Indoor Picocells

Within an office or large building environment, the reflections and absorptions of RF power by walls and floors, the strongly rectangular orientation of the obstacles and the slow movement of mobiles and scatterers (read people) make quite a different propagation environment from the two outdoor channels we reviewed earlier.

Path Loss

Multifloor buildings encourage a three-dimensional system organization in which cells are associated with individual floors or groups of floors, in contrast to the conventional planar organization of outdoor cellular. These sketches suggest why.

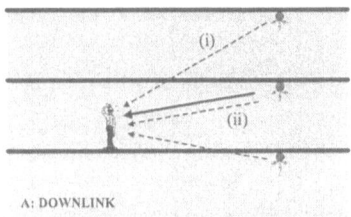 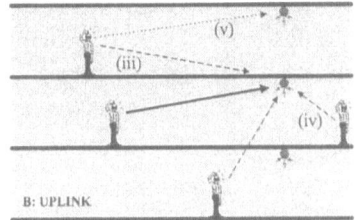

Interference scenarios for DS-CDMA in a multifloor environment. Solid lines: desired signal. Dashed lines: interference. Small roman numerals identify different classes of interference. From [**Butt98**] (© 1998 IEEE)

The attenuation between cochannel cells depends more on number of floors traversed than on distance, which you recall was the dominant **path loss** factor in outdoor systems, with their planar organization. To quantify the attenuation, [**Stee92**] cites a model by [**Keen90**] which gave a good fit to measurements:

$$\text{path\_loss} = L(v) + 20 \cdot \log(d) + n_f \cdot a_f + n_w \cdot a_w$$

where

n_f, n_w	number of floors, walls, crossed
a_f, a_w	attenuation per floor, wall
d	3 dimensional straight line distance (so inverse square)
$L(v)$	log-normal, mean between 32 and 38, variance v (standard deviation 3-4 dB)

Values of the attenuation coefficients depend strongly on the construction of individual buildings, especially factors such as thickness of the floors and the amount of reinforcing steel. This model specifies inverse square law, or a **path loss exponent** of 2. However, [**Bult87**], [**Bult89a**] give a path loss exponent along corridors of 1 at 1.7 GHz and 1-3 at 860 MHz. [**Todd92**] gives exponents of 3.15 for obstructed paths and 1.8 for LOS paths. Prediction for specific sites was examined in [**Seid94**], with coverage region prediction in [**Panj96**].

Shadowing

Some interesting results on shadowing in indoor propagation have been reported in
[**Butt97**]. During a program of measurements on typical office buildings, the authors
discovered strong floor-to-floor correlation of shadowing, which they attributed to the fact
that the footprint of the major structural elements of an office tower is very similar from one
floor to another. Based on that observation, they developed new guidelines for positioning the
base stations on alternate floors to maximize the indoor system capacity [**Butt98**].

Scenario	I	II	III
Users/Floor	Downlink Outage Probability		
10	0.0013	0.0049	0.0008
20	0.0031	0.0096	0.0012
30	0.0052	0.0152	0.0023
	Uplink Outage Probability		
10	0.0237	0.0224	0.0380
20	0.0447	0.0469	0.0824
30	0.0679	0.0691	0.1192

The table above shows three scenarios for base station placement by floor: antennas at the
same location on every floor; antennas at opposite ends of alternate floors; and antennas at the
same location on every floor, but different frequency bands used on alternate floors. Outage
is the fraction of floor space in which the bit error rate exceeds 10^{-2}. These results shown that
it is preferable to locate antennas at the same locations on every floor. In this case, the
desired signal and interference grow weaker and stronger together, thereby keeping variations
in carrier to interference ratio relatively small; in contrast, the second scenario produces much
more variation, resulting in one end of each floor with poor C/I. These results apply to fixed
wireless access as much as to pedestrian use.

Power Delay Profile

For obvious reasons, the rms delay spread is much smaller within buildings than outdoors.
[**Hash93**] gives typical values of 20-30 ns, and up to 125 ns for very large buildings.
Measurements presented in [**Jans96**] have a median value of rms delay spread between 9 ns
and 15 ns. The power delay profile is approximately exponential.

8. DIFFERENCES BETWEEN MOBILE AND BASE CORRELATIONS

Except for part of the analysis of Doppler spectrum in **Section 5.1**, we have so far considered only isotropic antennas and isotropic scattering. This is a considerable oversight, since the effects of antenna directionality and diversity are fundamental to good mobile system design. Whether achieved by means of "smart" antenna arrays, horns or other methods of beam forming, they offer the prospect of greatly reduced damage from fading, as well as space division multiple access - i.e., more than one user occupying the same time and frequency. This is one of the most active areas of research and development today.

These topics are the subject of other texts in this series: *Detection and Diversity* and *Smart Antenna Arrays for Mobile Communication*. However, your ability to deal with them rests on a clear understanding of channel behaviour. This section launches you into it.

The main results developed here are:

* spatial correlation (temporal correlation if the antenna is moving) and its dependence on the angular location of scatterers with respect to the direction of displacement and on the angular dispersion, or beamwidth;

* differences between typical spatial correlations at the mobile and at the base station.

8.1 Directionality at the Mobile - Doppler Spread

When we obtained the Doppler spectrum in **Section 5.1**, we were primarily concerned with scatterer power that was isotropic in azimuth - i.e., the scatterers were uniformly distributed around the mobile. However, nonisotropic distributions, whether they arise through a nonuniform distribution of physical scatterers or through antenna directionality, have the potential to allow new strategies in diversity and fading mitigation. Apart from its inherent interest, this investigation of directionality at the mobile sets the stage for the very important discussion of directionality at the base station in Section 8.3.

8.2 Directionality at the Mobile - Delay Spread

Directional antennas at the mobile can also reduce the delay spread - when pointed in the right direction, that is. This can make a real difference in a system, since the complexity of equalizers or other compensation methods represents a significant portion of both development and manufactured hardware costs.

8.3 Angular Dispersion and Directionality at the Base

The correlation of complex gain for various antenna displacements is strikingly different at the base station than it is at the mobile. It depends on the direction of the displacement and on the angular spread of the signals arriving from the scatterers surrounding the mobile. Understanding this phenomenon is critical to design of smart antenna arrays.

8.1 Directionality at the Mobile - Doppler Spread

When we obtained the Doppler spectrum in **Section 5.1**, we were primarily concerned with scatterer power that was isotropic in azimuth - that is, the scatterers were uniformly distributed around the mobile.. However, nonisotropic distributions, whether they arise through a nonuniform distribution of physical scatterers or through antenna directionality, have the potential to enable interesting new strategies in diversity and mitigation of fading.

Directionality at the mobile has not found much use to date, largely because of the cost and size of the antenna structure. In addition, cell phones can be held in any orientation, which demands rapid adaptation. Nevertheless, the move to higher frequencies, coupled with recent developments in adaptation algorithms and specific technologies such as low cost DSP and MMICs, suggests that smaller, cheaper antenna arrays are possible. Let's see what they might do for us.

Directionality Affects Fade Rate and Frequency Offset

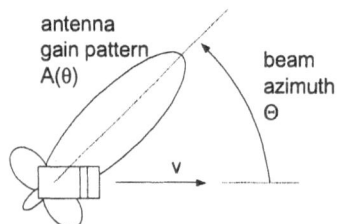

The mobile, seen from above, is moving right at speed v. It is superimposed on the amplitude gain pattern $A(\theta)$ of its antenna.

As in **Section 5.1**, we denote the incoming signal power in differential angle $d\theta$ at θ by $P(\theta)d\theta$, and denote the antenna amplitude gain pattern by $A(\theta)$, both referenced to the direction of motion, defined by $\theta=0$. With a slight modification of (5.1.4), we have the Doppler spectrum of received power as

$$S_g(\nu) = \frac{A(\theta)^2 \cdot P(\theta) + A(-\theta)^2 \cdot P(-\theta)}{f_D \sqrt{1 - \cos(\theta)^2}} \qquad (8.1.1)$$

where ν is the frequency and $\theta = arccos(\nu/f_D)$. In this section, we'll consider only isotropic scattering

$$P(\theta) = \frac{\sigma_g^2}{2 \cdot \pi} \qquad (8.1.2)$$

and a directional antenna, so that

$$S_g(\nu) = \frac{\sigma_g^2}{2 \cdot \pi} \cdot \frac{A(\theta)^2 + A(-\theta)^2}{f_D \sqrt{1 - \cos(\theta)^2}} \qquad (8.1.3)$$

A very simple model for the directional antenna gives it a rectangular pattern with a given beamwidth Ω and direction Θ. The plot below shows five beams.

$$\text{close}(\alpha) := \alpha - 2 \cdot \pi \cdot \text{floor}\left(\frac{\alpha}{2 \cdot \pi} + \frac{1}{2}\right) \qquad \text{(fixes mod } 2\pi \text{ problems)}$$

$$A(\theta, \Theta, \Omega) := \left(\left| \text{close}(\theta - \Theta) \right| \le \frac{\Omega}{2}\right) \qquad \text{antenna pattern, direction } \Theta, \text{ width } \Omega \qquad (8.1.4)$$

$$\Omega := 0.5 \qquad\qquad \theta := -\frac{255}{256} \cdot \pi, -\frac{253}{256} \cdot \pi \, .. \, \pi + \frac{\pi}{256}$$

$$\Theta_1 := 0 \qquad \Theta_2 := \frac{\pi}{2} \qquad \Theta_3 := \pi \qquad \Theta_4 := -\frac{\pi}{4} \qquad \Theta_5 := 0.85 \cdot \pi$$

$\dfrac{A\,\theta, \Theta_1, \Omega}{}$

$\dfrac{A\,\theta, \Theta_2, \Omega}{}$

$A\,\theta, \Theta_3, \Omega$

$A\,\theta, \Theta_4, \Omega$

$A\,\theta, \Theta_5, \Omega$

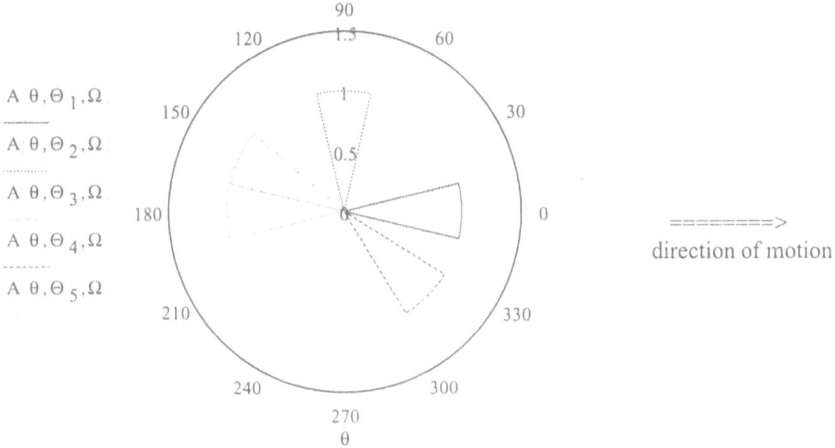

direction of motion

Idealized Antenna Patterns

The beams select specific azimuthal ranges, and therefore specific ranges of Doppler shift. We can see the Doppler spectra corresponding to these beams by plotting (8.1.3), using the parametric link $v = f_D \cos(\theta)$.

$$\sigma_g := 1 \qquad f_D := 1$$

$$S_g(\theta, \Theta, \Omega) := \frac{\sigma_g^2}{2 \cdot \pi} \cdot \frac{A(\theta, \Theta, \Omega)^2 + A(-\theta, \Theta, \Omega)^2}{f_D \sqrt{1 - \cos(\theta)^2}} \qquad v(\theta) := f_D \cdot \cos(\theta)$$

$$(8.1.5)$$

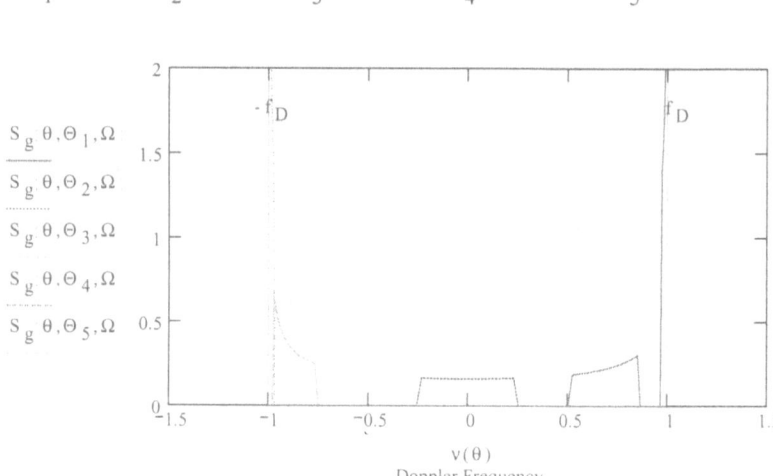

Doppler Spectra, Directional Antennas

We can draw several fascinating inferences from this plot and the assumptions that led up to it:

* Because the antennas interact with different sets of scatterers, they have statistically independent complex gains (or impulse responses, more generally). As a result, we obtain diversity reception. The full significance of diversity may not become apparent until you study topics covered in the texts *Detection and Diversity* and *Coding for Mobile Communications*, but you can appreciate its importance from the fact that not all beams are likely to fade simultaneously. That (almost) ensures at least some signal at any time.

* Partially offsetting the diversity improvement is the fact that each antenna intercepts a fraction $\Omega/2\pi$ of the total scatterer power available. This makes the average SNR per antenna less than that of an isotropic antenna by the same factor. Note that all antennas pick up the same average power because they subtend the same angle, even though their spectra are of different widths.

* Because the spectra are not necessarily centred on 0 Hz, their complex gains have a net tendency to rotate. If the detection algorithm is sensitive to frequency offset, the received signal should be demodulated; i.e., shifted to 0 Hz by multiplying by $\exp(-j2\pi f_D\cos(\Theta)t)$. Of course, we must estimate f_D in order to do so - but at least we know Θ from the orientation of the antenna.

* The fade rate and level crossing rate of the signals from the antennas are proportional to their rms bandwidths, so they are much less than the values experienced by an isotropic antenna, in which the Doppler spectrum runs from $-f_D$ to f_D. Since both the error floor and the difficulties in tracking the gain for adaptation increase with fading rate, this reduction is significant, especially as systems move to higher carrier frequencies, where f_D is greater. More about this in Section 3.1.1 of [**Jake74**]. Incidentally, an equivalent statement is that the coherence distance is increased from the value $x_c=\lambda/\pi\sqrt{2}$ that we obtain from (5.1.17) with $x_c=vT_c$.

* The widths of the spectra depend on the direction Θ of the beam, as well as the beamwidth Ω. Antennas pointing directly ahead or behind (i.e., along the direction of motion) experience very slow fading, since their power spectra are very narrow. Antennas pointing transverse (i.e. 90 degrees) to the direction of motion have much faster fading (though still less than that of an isotropic antenna). In spatial terms, the complex gain decorrelates more quickly with antenna displacements that are 90 degrees off the line to the illuminated scatterers than displacements directly toward or away from the scatterers.

Details: Coherence Time and Level Crossing Rate

We can expand that last observation and calculate the coherence time and distance and the level crossing rate. We know from **(5.1.16)** of Section 5.1 that reasonable definitions of coherence time and coherence distance are, respectively,

$$T_c = \frac{1}{2 \cdot \pi \cdot v_{rms}} \qquad\qquad x_c = \frac{v}{2 \cdot \pi \cdot v_{rms}} \qquad\qquad (8.1.6)$$

Also, from **(6.2.11)** of Section 6.2, the level crossing rate is

$$n_f = 2 \cdot \sqrt{\pi} \cdot v_{rms} \cdot R \cdot e^{-R^2} \qquad\qquad (8.1.7)$$

Evidently, a calculation of rms Doppler spread v_{rms} will give us all of these quantities. To obtain it, recall that the nth moment of Doppler spread is

$$v^{(n)} = \frac{2 \cdot \pi}{\Omega} \cdot \frac{1}{\sigma_g^2} \cdot \int_{v_{min}}^{v_{max}} v^n \cdot S_g(v) \, dv \qquad\qquad (8.1.8)$$

where the leading factors normalize the power spectrum, and that

$$v_{rms}^2 = v^{(2)} - \left[v^{(1)} \right]^2 \qquad\qquad (8.1.9)$$

To evaluate the moments, we go back to **(5.1.1), (5.1.2) and (5.1.4)** of Section 5.1, reproduced here as

$$v = f_D \cdot \cos(\theta) \qquad\qquad dv = -f_D \cdot \sqrt{1 - \left(\frac{v}{f_D} \right)^2} \cdot d\theta$$

$$S_g(v) \cdot dv = \left(A(\theta)^2 \cdot P(\theta) + A(-\theta)^2 \cdot P(-\theta) \right) \cdot \frac{dv}{f_D \cdot \sqrt{1 - \left(\frac{v}{f_D} \right)^2}} \qquad\qquad (8.1.10)$$

Rewriting (8.1.8) and substituting (8.1.2), we have

$$v^{(n)} = \frac{1}{\Omega} \cdot \int_{\Theta - \frac{\Omega}{2}}^{\Theta + \frac{\Omega}{2}} \left(f_D \cdot \cos(\theta) \right)^n d\theta \qquad (8.1.11)$$

This is easily evaluated (especially with the help of the Mathcad symbolic processor) as

the mean:

$$v^{(1)} = \frac{2}{\Omega} \cdot f_D \cdot \cos(\Theta) \cdot \sin\left(\frac{\Omega}{2}\right) \qquad (8.1.12)$$

and the second moment:

$$v^{(2)} = f_D^2 \cdot \frac{\sin(\Omega) \cdot \cos(2 \cdot \Theta) + \Omega}{2 \cdot \Omega} \qquad (8.1.13)$$

Therefore the rms bandwidth, normalized by f_D, is

$$v_{rms\_n}(\Theta, \Omega) := \sqrt{\frac{\sin(\Omega) \cdot \cos(2 \cdot \Theta) + \Omega}{2 \cdot \Omega} - \frac{4}{\Omega^2} \cdot \cos(\Theta)^2 \cdot \sin\left(\frac{\Omega}{2}\right)^2} \qquad (8.1.14)$$

and, for example, the coherence distance from (8.1.6) is, in wavelengths,

$$x_{c\_n}(\Theta, \Omega) := \frac{1}{2 \cdot \pi \cdot v_{rms\_n}(\Theta, \Omega)} \qquad (8.1.15)$$

Plots tell a lot. Check the variation with antenna orientation for four different beamwidths:

$$\Theta := 0, \frac{\pi}{32} .. \frac{\pi}{2} \qquad \Omega_1 := 0.1 \qquad \Omega_2 := 0.2 \qquad \Omega_3 := \frac{\pi}{4} \qquad \Omega_4 := \frac{\pi}{2}$$

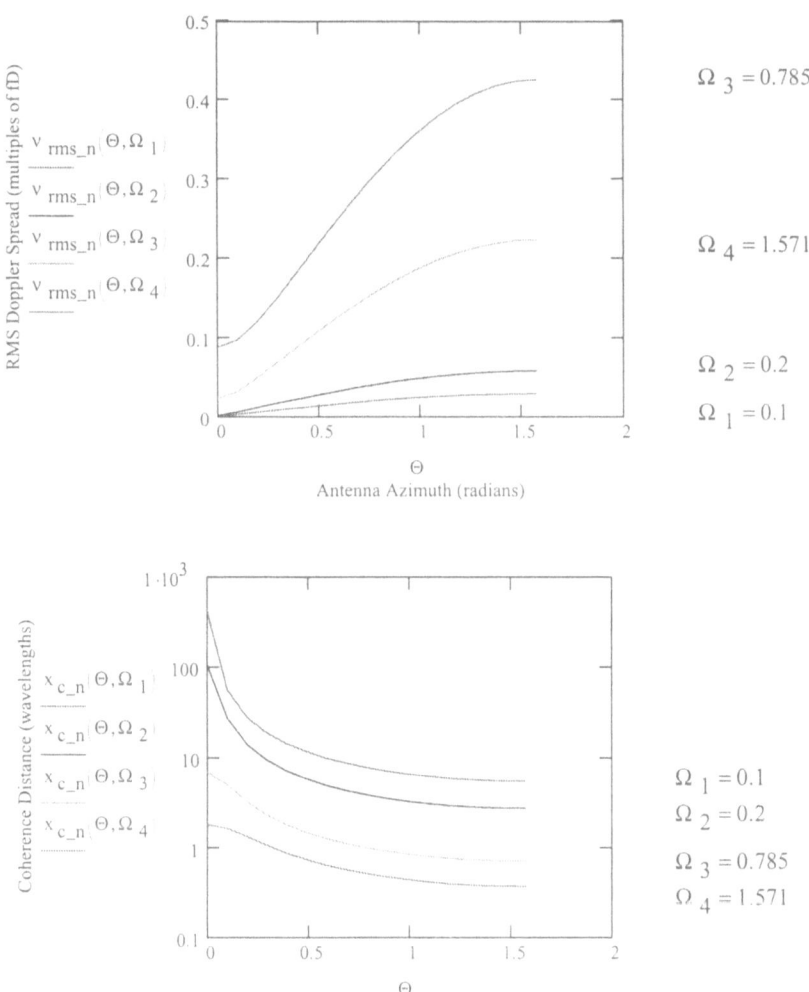

We see dramatic differences in both coherence time and distance as the beam direction changes from directly ahead ($\Theta=0$) to transverse ($\Theta=\pi/2$). For a narrow 0.1 radian beamwidth, the coherence distance ranges from 400 wavelengths (!) for scatterers directly ahead to just 5.5 wavelengths for transverse scatterers. For a more easily constructed 45 degree beamwidth ($\Omega=p/4$), the coherence distance varies from 7.5 wavelengths to 0.7 wavelengths.

A series expansion in beamwidth gives more insight than the exact (8.1.14). For small beamwidths, the normalized fading bandwidth can be written

$$v_{rms\_c} = \frac{\Omega}{2 \cdot \sqrt{3}} \cdot \left| \sin(\Theta) \right| \quad \text{approx, } \Omega \ll 1 \tag{8.1.16}$$

As the graph below shows, this approximation is quite good for beamwidths less than about 45 degrees, except near $\Theta=0$ (directly ahead). This latter value is obtained by an alternative series expansion as

$$v_{rms\_c} = \frac{\sqrt{5}}{60} \cdot \Omega \qquad \text{approx, } \Omega<<1 \text{ and } \Theta=0 \qquad (8.1.17)$$

$$\Theta := 0, 0.01 .. \frac{\pi}{2}$$

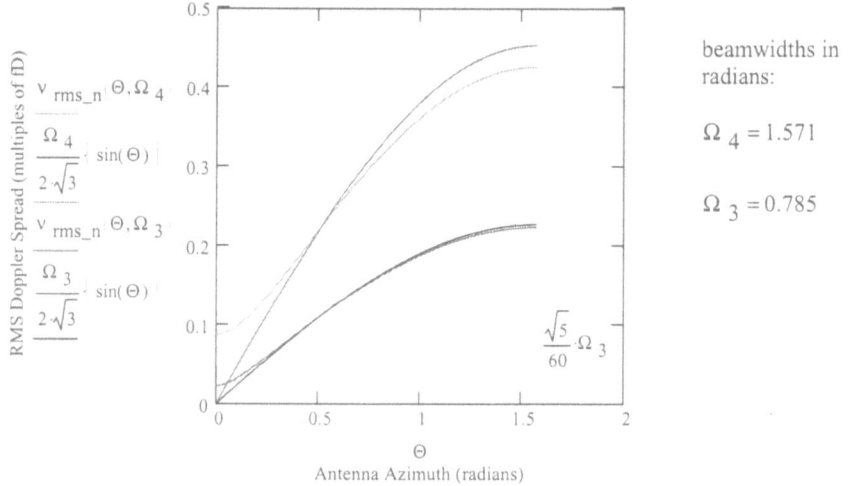

RMS Doppler Spread (multiples of fD)

$v_{rms\_n}(\Theta, \Omega_4)$

$\frac{\Omega_4}{2\sqrt{3}} \cdot \sin(\Theta)$

$v_{rms\_n}(\Theta, \Omega_3)$

$\frac{\Omega_3}{2\sqrt{3}} \cdot \sin(\Theta)$

$\frac{\sqrt{5}}{60} \cdot \Omega_3$

Θ
Antenna Azimuth (radians)

beamwidths in radians:

$\Omega_4 = 1.571$

$\Omega_3 = 0.785$

RMS Doppler Spread and Approximations

Evidently, a reasonable approximation in the case of small beamwidth is

$$v_{rms\_c} = \Omega \cdot \max\left(\frac{\sqrt{5}}{60}, \frac{|\sin(\Theta)|}{2\cdot\sqrt{3}}\right) \qquad (8.1.18)$$

We now have results for three quantities of interest: coherence time, coherence distance and level crossing rate, all in terms of the antenna direction and beamwidth. They all rest on the rms bandwidth of the fading process picked up by the antenna. Here the series approximations (8.1.16) and (8.1.17) are probably more useful than the exact expression (8.1.14). Why? Because they give useful rules of thumb - rms Doppler is roughly proportional to the product of beamwidth and sine of the azimuth - and in any case, the exact expression is based on idealized (i.e., approximate) models of scatterers and beam, so its claim to legitimacy is somewhat muted.

Before we leave this topic, recall that there is little difference between the signal produced by a single directional antenna in uniformly distributed scatterers and that produced by an omnidirectional antenna when there is a single cluster of scatterers at some azimuth. In this latter case, we also see the combination of equivalent frequency offset and reduced Doppler spread (increased coherence distance), both dependent on direction of motion with respect to azimuthal location of the scatterer cluster.

8.2 Directionality at the Mobile - Delay Spread

In **Section 8.1**, we established that directional antennas at the mobile can considerably reduce the **Doppler spread** experienced by the individual antenna signals, with correspondingly increased coherence distances and reduced level crossing rates. We also found a strong dependence on direction of the antenna.

In this section, we will look at the other domain - **delay spread** - and assess the effect of antenna directionality. Again, we will see reduction of the amount of spread and an orientation dependence. Reduction of delay spread is very significant if it can simplify, or even eliminate, the equalizer. These devices represent a significant fraction of development costs and a major sink of DSP cycles - i.e. unit manufactured cost.

We'll use a very simple model for delay. Assume that every path between mobile and base bounces from exactly one scatterer. Then the locus of scatterer locations with the same delay is an ellipse, with the mobile and base as foci. With a little rearrangement of the usual equation for an ellipse, we can express it in polar coordinates r,θ with the origin at the mobile as

$$r(\theta, d_m, d_x) := \frac{1}{2} \cdot d_x \cdot \frac{(d_x + 2 \cdot d_m)}{(d_m \cdot \cos(\theta) + d_m + d_x)} \tag{8.2.1}$$

where d_m is the minimum distance between base and mobile (i.e., the distance between foci) and d_x is the excess distance; that is, the difference between the base to mobile distance with a bounce off the ellipse and the minimum distance. The excess distance, when divided by the speed of light, corresponds directly to the delay axis τ in the impulse response.

$\theta := -\pi, -.99 \cdot \pi .. \pi$

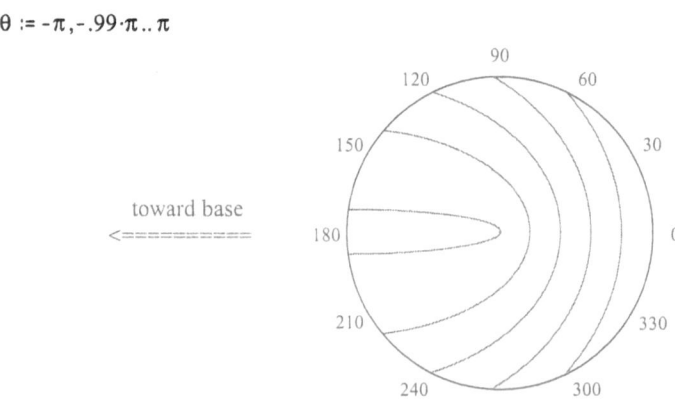

The situation as represented in this simple model can be visualized by interpretation of the graph, which represents the neighbourhood of significant scatterers around the mobile. The concentric green lines represent contours of equal power (unfortunately equispaced), since the average power decreases with distance of the scatterer from the mobile, down to zero power at the outside of the graph. The red ellipse segments are "iso-delay" curves, with the inner one corresponding to a very small delay. The green radial lines suggest antenna sectors.

Examination of the graph shows that antenna direction makes a very large difference to delay spread. If the antenna points toward the base, it illuminates only those scatterers with low delay. Conversely, if the antenna points away from the base, it illuminates scatterers covering the whole range of delays, and it consequently experiences a large delay spread. The difference between them is significant - and possibly enough to spell the difference between a simple receiver and one that requires an equalizer. By the way, remember that all antennas of the same beamwidth intercept the same power, assuming isotropic scattering, regardless of its delay distribution.

How does this variation in delay spread relate to the antenna-induced variation in Doppler spread we saw in **Section 8.1**? Not at all. To repeat the theme from **Section 3.3**, delay spread and Doppler spread are two separate phenomena. In the graph above, the mobile could be moving in any direction. The antenna facing in that direction would experience the maximum Doppler shift and the minimum Doppler spread, irrespective of the angular location of the base. The combination makes some interesting **scattering functions**.

8.3 Angular Dispersion and Directionality at the Base

Diversity at the base station is fundamental to good mobile communications design. But how far apart should the antennas be spaced to obtain adequate diversity? Here we see a striking difference between behaviour at the base station and the mobile. You recall from **Section 5.1** that, at a typical mobile with a single isotropic antenna and surrounded by scatterers, decorrelation of the complex gain occurs over distances as short as half a wavelength. In contrast, you will see in this section that, at a typical base station, decorrelation requires separations of tens of wavelengths or more. Another difference is a strong dependence on direction of the separation with respect to the angular location of the mobile.

You may not have expected these differences between mobile and base, since the channel is reciprocal in the electromagnetic sense. However, they become clear if you consider the different geometries: the mobile is surrounded by scatterers and usually employs an isotropic antenna (though see **Section 8.1**), whereas the base is usually located where it is not troubled by local scatterers (e.g., on a tower) and it receives signals from the mobile over a narrow angular range that is determined by the size of the mobile's scattering neighbourhood and its distance from the base. In addition, the base frequently uses a directional antenna, such as a 120 degree or 60 degree sectoral antenna.

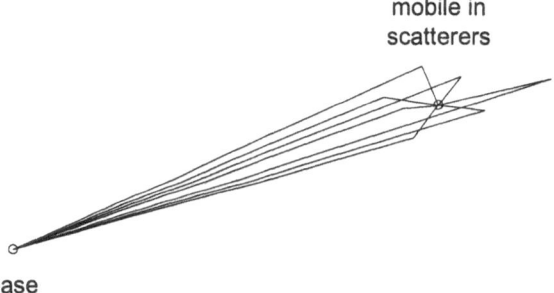

mobile in
scatterers

base

In this section, we will build a body of theory you need for diversity antenna design and for your study of smart, adaptive antenna arrays. We will start with a simple semi-quantitative argument that exposes the basic issues, then adapt the results of our analysis of directionality at the mobile in **Section 8.1** to show the dependence of coherence distance on angular dispersion. Both discussions keep mathematics to a minimum and try to present the phenomena in an intuitive way.. Finally, we will obtain the autocorrelation function of complex gain - the fundamental issue in diversity design - as a function of spatial separation at the base, using an idealized model.

Simple Argument

Consider the simplified configuration in the sketch below. The mobile (seen from above) is a distance d from the base, and there are only two scatterers, separated by a distance s that is measured transverse to the axis, or line connecting base and mobile. Since decorrelation is a result of changing phase angles among the paths, we ask what is the differential distance and phase between the two paths if the base antenna is moved a distance x, also measured transverse to the axis.

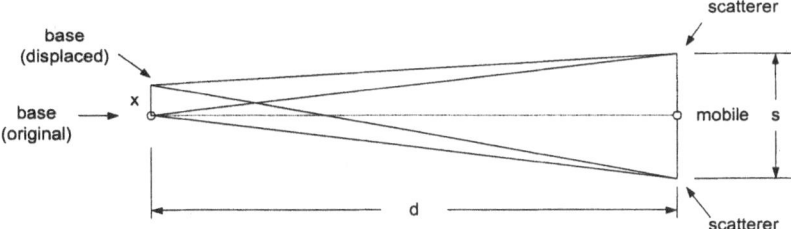

A straightforward analysis based on right triangles gives the differential distance as

$$\sqrt{d^2 + \left(\frac{s}{2} + x\right)^2} - \sqrt{d^2 + \left(\frac{s}{2} - x\right)^2} \equiv d \cdot \sqrt{1 + \frac{\left(x + \frac{s}{2}\right)^2}{d}} - d \cdot \sqrt{1 + \frac{\left(x - \frac{s}{2}\right)^2}{d}} \quad (8.3.1)$$

and series expansion of the square roots lets us approximate it as

$$d \cdot \left[1 + \frac{1}{2} \cdot \frac{\left(x + \frac{s}{2}\right)^2}{d}\right] - d \cdot \left[1 + \frac{1}{2} \cdot \frac{\left(x - \frac{s}{2}\right)^2}{d}\right] \equiv x \cdot \frac{s}{d} \quad \text{so the differential phase is } \frac{x}{\lambda} \frac{s}{d}$$

$$(8.3.2)$$

We can observe the principal phenomena in this simple result:

* The quantity s/d is the angular dispersion, analogous to beamwidth Ω in **Section 8.1**. We might have the scatterer diameter s as one or two hundred metres. In a large cell, where the distance d is large, we will therefore find very small values of angular dispersion - as low as 0.01 radian - so that x must be *many* wavelengths for a significant phase change, or significant decorrelation. Even in a microcellular environment, the angular dispersion is on the order of 0.1, and is unlikely to exceed 0.5 radian, so that *the required antenna separation at the base is still an order of magnitude greater than at the mobile.*

* From **(8.1.15) and (8.1.18)**, as well as (8.3.2) above, the coherence distance depends inversely on the angular dispersion (i.e., beamwidth Ω) for any orientation of mobile with respect to antenna displacement. Therefore we expect faster decorrelation, and consequently greater diversity, for mobiles that are closer to the base. Faraway mobiles appear as point sources, and are coherent in the displacement neighbourhood of the base station.

* Now generalize to the case of many scatterers and consider the impulse response. If the differential delay changes caused by base antenna displacement are small compared to the wavelength, they are negligible compared to the time scale of the modulation. Therefore the scatterers within each delay bin are fixed in individual amplitudes but sum to a different resultant as the differential delays cause phase changes of the carrier. We conclude from this that all base antenna locations in a neighbourhood of tens of wavelengths have the same power delay profile.

* We obtain the greatest change in differential distance with this transverse displacement of the base antenna. In contrast, displacements along the axis have *no* effect on differential delays, so that there is no possibility of diversity in that direction. In reality, of course, there are more than two scatterers and they are not confined to a straight line transverse to the axis, so the changes in differential path length are not strictly zero. However, they are very small, and there is very little decorrelation with antenna displacement along the axis. This means that if you have a linear array, as in the sketch below, don't expect much decorrelation in the endfire direction. Even broadside has slow decorrelation.

linear array at base

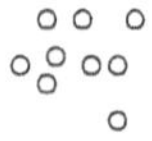

an endfire cluster

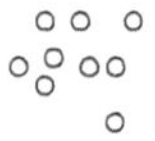

a broadside cluster

Coherence Distance

Our next step up in model sophistication is more quantitative: the coherence distance. We already obtained this quantity in **Section 8.1** in connection with directionality at the mobile. We can use those results directly if we are willing to believe that the set of paths between the base and the scatterers associated with a particular mobile has power that is distributed uniformly over a small range of azimuth. From **(8.1.15) and the approximations (8.1.16) and (8.1.17)**, the coherence distance in wavelengths is then

$$x_c = \frac{1}{2 \cdot \pi \cdot \Omega} \cdot \min\left(\left| \frac{2 \cdot \sqrt{3}}{\sin(\Theta)} \right|, \frac{60}{\sqrt{5}} \right) \qquad (8.3.3)$$

This is consistent with our observations on the simple model above: the coherence distance is inversely proportional to the angular dispersion Ω and it is greatest at $\Theta=0$; that is, the direction of the mobile.

Autocorrelation Function

In the discussion so far, we have seen the principal effects of angular location and angular dispersion of the mobile's scatterers on coherence at the base station - and we found them without having to do detailed mathematics. If we intend to do any real analysis of diversity or adaptive beam forming, though, we need the autocorrelation of complex gain with respect to antenna displacements at the base. And, unfortunately, that means the holiday is over. It will take a little work to get this result.

The sketch below shows the situation. We have a reference location for the base antenna, and the mobile is at some angular position Θ. As usual, the mobile is surrounded by a group of scatterers. We are interested in the correlation between the complex gain $g(0)$ at the reference position and the complex gain $g(x)$ at a distance x away.

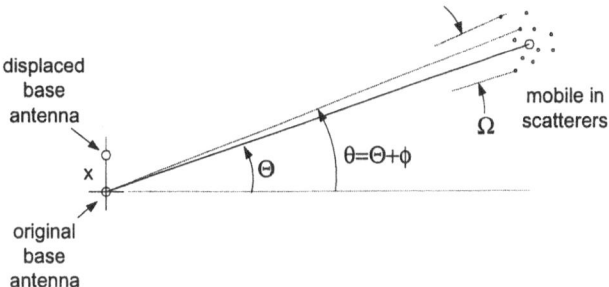

At any angle θ, the arrivals in a differential angle $d\theta$ sum to the complex amplitude $A(\theta)$. Then the complex gain at the reference position (the original base antenna) is just the sum of all scatterers

$$g(0) = \int_{-\pi}^{\pi} A(\theta) \, d\theta \tag{8.3.4}$$

At the new position, the path length for scatterers at azimuth θ has decreased by $x \cdot \sin(\theta)$, which introduces a corresponding phase change. In reality, this expression for path length change is an approximation that holds for displacements that are much smaller than the distance to the mobile (*not* the same thing as small angular dispersion). We owe it to Jakes, who used it for both mobile and base autocorrelations [**Jake74**], where it appears as a Doppler shift in his derivations. In any case, the resultant complex gain becomes

$$g(x) = \int_{-\pi}^{\pi} A(\theta) \cdot \exp(j \cdot \beta \cdot x \cdot \sin(\theta)) \, d\theta \tag{8.3.5}$$

where β is the wave number $2\pi/\lambda$. The autocorrelation function as a function of position is

$$R_g(x) = \frac{1}{2} \cdot E\left(g(x) \cdot \overline{g(0)}\right)$$

$$= \frac{1}{2} \cdot \int_{-\pi}^{\pi} \int_{-\pi}^{\pi} E\left(A(\theta) \cdot \overline{A(\alpha)}\right) \cdot \exp(j \cdot \beta \cdot x \cdot \sin(\theta)) \, d\theta \, d\alpha \tag{8.3.6}$$

Since the scatterers at different angles θ and α are independent (they are different scatterers), we have

$$R_g(x) = \int_{-\pi}^{\pi} P(\theta) \cdot \exp(j \cdot \beta \cdot x \cdot \sin(\theta)) \, d\theta \tag{8.3.7}$$

where $P(\theta)$ is the angular power density

$$P(\theta) = \frac{1}{2} \cdot E\left[(| A(\theta) |)^2 \right] \tag{8.3.8}$$

In one sense, (8.3.7) is simple, because it shows that the autocorrelation function is determined by the angular power density. On the other hand, we can make things arbitrarily difficult for ourselves by trying to solve it exactly. Recall that, even in the easy case of a uniform $P(\theta)$ over $[-\pi,\pi)$, the autocorrelation function is the Bessel function $J_0(\beta x)$, and things can only get worse from there. In reality, though, there is little need for an exact solution, since even our starting point was an approximate model of the true propagation environment (and if we really cared, we could evaluate (8.3.7) numerically, at the cost of insight).

So we'll take a simpler tack. Represent the azimuth θ of the scatterers as a small increment ϕ on the angular location Θ of the mobile, so that $\theta = \phi + \Theta$ (see the sketch above). Next, expand the sine in the exponential in (8.3.7) trigonometrically, then in a series in ϕ

$$\sin(\phi + \Theta) = \sin(\Theta) \cdot \cos(\phi) + \cos(\Theta) \cdot \sin(\phi)$$

$$= \sin(\Theta) + \cos(\Theta) \cdot \phi - \frac{1}{2} \cdot \sin(\Theta) \cdot \phi^2 + O(\phi^3) \tag{8.3.9}$$

The angular dispersion Ω is usually small: $\Omega = 0.4$ would be a very large value, seen only in a microcellular system, and an order of magnitude smaller is more typical of a macrocell. Consequently, we can truncate the series (8.3.9) to the linear term in ϕ. Substitution of (8.3.9) into (8.3.7) gives

$$R_g(x) = e^{j \cdot \beta \cdot x \cdot \sin(\Theta)} \cdot \int_{-\frac{\Omega}{2}}^{\frac{\Omega}{2}} P(\phi + \Theta) \cdot e^{j \cdot \beta \cdot x \cdot \cos(\Theta) \cdot \phi} \, d\phi$$

$$= e^{j \cdot \beta \cdot x \cdot \sin(\Theta)} \cdot p\left(\frac{x}{\lambda} \cdot \cos(\Theta) \right) \tag{8.3.10}$$

Fascinating! The autocorrelation function is proportional to the inverse Fourier transform p of the angular power density P, dilated by the factor $1/\cos(\Theta)$.

Here's a quick inference from (8.3.10). Represent the displacement of the base station antenna as a component $x\sin(\Theta)$ along the axis between base and mobile (the endfire direction) and a component $x\cos(\Theta)$ transverse to the axis (the broadside direction). The endfire component affects only the phase of the autocorrelation, not the magnitude, so that it has no effect on diversity. It is just the broadside component that provides diversity. Of course, these statements are valid only to the validity of ignoring second and higher order terms in the series (8.3.9).

Autocorrelation Functions for Three Scatterer Models

We'll look at three candidates for angular power density $P(\theta)$: a uniform distribution, a ring of scatterers and a disc of scatterers. They and their corresponding autocorrelation functions are shown together in a graph following the discussion.

To simplify things, we'll express distance in wavelengths. That's equivalent to setting $\lambda := 1$. Also, we'll normalize to unit power, or $\sigma_g := 1$. Finally, we define

$$\mathrm{sinc}(x) := \mathrm{if}\left(x=0,1,\frac{\sin(\pi \cdot x)}{\pi \cdot x}\right)$$

The *first candidate* for angular power density is a uniform distribution over the beamwidth Ω. We used it in **Section 8.1** as a simple model for the signal environment at a mobile with a directional antenna in isotropic scattering, and it was also employed in [**Salz94**] as a base station model. This gives

$$P_{\mathrm{unif}}(\phi,\Theta,\Omega) := \begin{vmatrix} \dfrac{\sigma_g^{\,2}}{\Omega} & \text{if } |\phi - \Theta| \le \dfrac{\Omega}{2} \\ 0 & \text{otherwise} \end{vmatrix} \tag{8.3.11}$$

so that Fourier inversion produces the familiar sinc function:

$$R_{\mathrm{unif}}(x,\Theta,\Omega) := \sigma_g^{\,2} \cdot e^{\,j \cdot 2 \cdot \pi \cdot \frac{x}{\lambda} \cdot \sin(\Theta)} \cdot \mathrm{sinc}\left(\frac{x}{\lambda} \cdot \cos(\Theta) \cdot \Omega\right) \tag{8.3.12}$$

As a *second candidate*, you may feel that a more realistic model for angular power density would account for the clustering of scatterers around the mobile. Referring to the sketch above, let us assume that the scatterers are distributed uniformly on the circumference of a circle of radius $\Omega/2$ centred on Θ. This is the well-known "ring of scatterers" model, originally proposed by Jakes [**Jake74**]. A little geometry gives

$$P_{\mathrm{ros}}(\phi,\Theta,\Omega) := \sigma_g^{\,2} \cdot \frac{2}{\pi \cdot \Omega} \cdot \frac{1}{\sqrt{1 - \left(2 \cdot \dfrac{\phi}{\Omega}\right)^2}} \quad \text{for} \quad |\phi| \le \frac{\Omega}{2} \tag{8.3.13}$$

It has the same shape as the "U-shaped spectrum" of **Section 5.1**, so its transform is the Bessel function. From (8.3.10), its autocorrelation function is

$$R_{ros}(x,\Theta,\Omega) := \sigma_g^2 \cdot e^{j \cdot 2 \cdot \pi \cdot \frac{x}{\lambda} \cdot \sin(\Theta)} \cdot J0\left(\pi \cdot \Omega \cdot \frac{x}{\lambda} \cdot \cos(\Theta)\right) \qquad (8.3.14)$$

For our *third candidate*, we take the scatterers to be uniformly distributed with equal power on a disc of radius $\Omega/2$ centred on Θ. To obtain the angular power density from the disc, we need the area in the disc that lies in $d\theta$ at θ. As you can imagine from the first part of the sketch below, this is a challenging computation in itself.

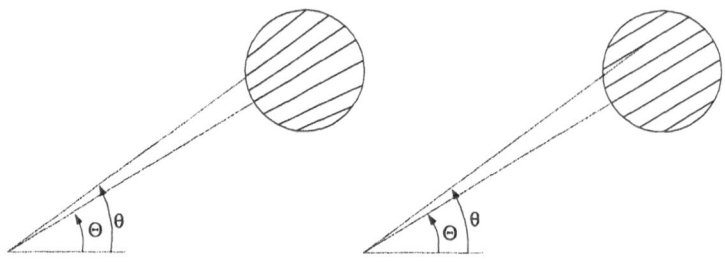

But why bother with this computation? We are assuming relatively small angular dispersion, so we might as well also assume that the differential area is a linear strip, rather than a wedge, as shown in the second part of the sketch. That makes an easy power density:

$$P_{dos}(\phi,\Theta,\Omega) := \sigma_g^2 \cdot \frac{8}{\pi \cdot \Omega^2} \cdot \sqrt{\left(\frac{\Omega}{2}\right)^2 - \phi^2} \qquad \text{for } |\phi| \le \frac{\Omega}{2} \qquad (8.3.15)$$

where the leading factors make its area σ_g^2. Unfortunately, a closed form for the inverse transform is not available. At this point, we could substitute (8.3.15) into (8.3.10) and perform the integration numerically - easy! Of course, if we are willing to perform a numerical integration, we might as well substitute (8.3.15) into the more accurate (8.3.7) and be done with it - but we'll continue with (8.3.10), in order to compare the three models on an equal footing.

Instead of continuing with (8.3.15), we'll employ an alternative calculation of the spatial autocorrelation R_{dos}. Consider the disc of scatterers as a set of concentric rings of scatterers. Then we can simply average R_{ros} frome (8.3.14) over the radius, with a weighting that reflects the increasing area at larger values of radius.

$$R_{dos}(x,\Theta,\Omega) := 2 \cdot \left(\frac{2}{\Omega}\right)^2 \cdot \int_0^{\frac{\Omega}{2}} \omega \cdot R_{ros}(x,\Theta,2\cdot\omega)\, d\omega \qquad (8.3.16)$$

The reason for using this form is that it opens the way to other radially dependent models in which the power of scatterers in a given ring is determined by some path loss law (see **Section 1**). You can take this idea the rest of the way yourself if you need it.

We can compare the three models graphically. First, the angular power densities. Assume that the mobile is at $\Theta = 0$ (the broadside direction) and that the angular dispersion is relatively large, $\Omega = 0.1$. Then plot the angular power densities (8.3.11), (8.3.13) and (8.3.15) for a specific parameter choice.

$$\Theta := 0 \qquad \Omega := 0.1 \qquad \phi := -\frac{\pi}{8}, -0.99 \cdot \frac{\pi}{8} .. \frac{\pi}{8}$$

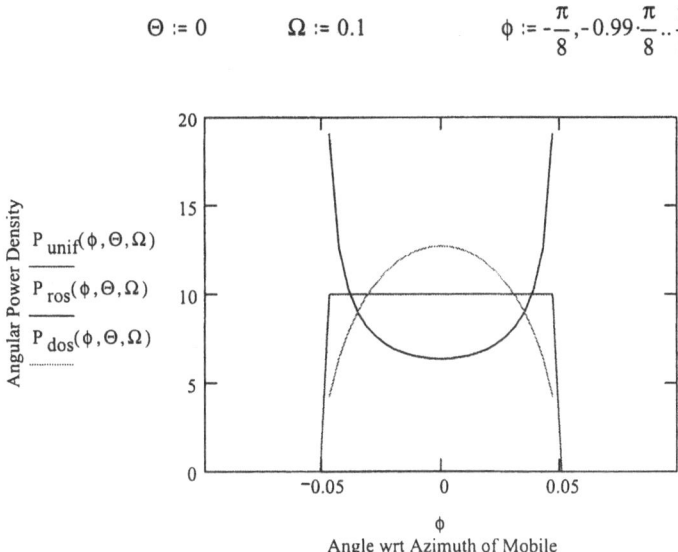

From basic Fourier theory, we see that the disc of scatterers should give a wider main lobe (greater coherence distance) than the uniform distribution, but its decay will be somewhat faster because the step discontinuity is softened. The ring of scatterers should have the slowest decay, because of its singularity.

Next, we'll look at the corresponding autocorrelation functions in the broadside direction $\Theta = 0$ (endfire just changes the phase). The plot below shows that the all three angular power densities allow complete decorrelation in 8 to 12 wavelengths. In fact, much of the diversity gain can be obtained with correlation coefficients as high as 0.5 (as we will see in the text *Detection and Diversity*), so it appears that a separation of 5λ to 7λ is sufficient. However, this plot is for angle spread $\Omega = 0.1$ or $\Omega = 5.73 \cdot \deg$ a fairly large value. From Fourier theory, and consistent with our conclusions at the end of **Section 8.1**, correlation distances scale inversely with the angular dispersion Ω. Therefore $\Omega = 0.01$ requires antenna separations of 50λ or more just for decorrelation to the 0.5 level. The DOS model drops even more slowly. A rule of thumb in base station design is a 20λ separation for diversity. At 900 MHz, this is about 7 metres - a large structure for a roof or tower! Fortunately, the shorter wavelength of PCS bands at 1.9 GHz will reduce it by a factor of 2.

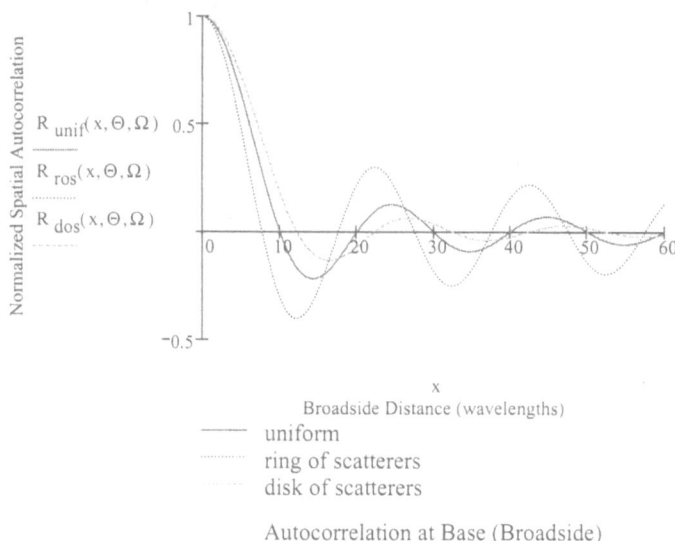

$R_{unif}(x,\Theta,\Omega)$ —— 0.5
$R_{ros}(x,\Theta,\Omega)$
$R_{dos}(x,\Theta,\Omega)$

Normalized Spatial Autocorrelation

x
Broadside Distance (wavelengths)
—— uniform
········· ring of scatterers
······ disk of scatterers

Autocorrelation at Base (Broadside)

A generalization of the simple angular scattering models we have just considered places the main contribution from the cluster in an environment of isotropic low-level clutter. [**Kalk97**] follows just this approach, modeling the angular power density as a strong uniform distribution of width Ω, just as above, but on a low-level pedestal of width 2π. As you might expect, the pedestal produces somewhat faster decorrelation.

Finally, note that all of these models - uniform, ring of scatterers and disc of scatterers - are very idealized. They are represent notions of how an average over many mobile locations might look. In any specific location, however, they might look quite different. In particular, the question of whether the received power is dominated by contributions from a few point scatterers is critical, since the answer tends to direct our thinking toward diversity, as above, or to eigenmethods that attempt to resolve the individual contributions and combine them coherently, thereby avoiding fading. These questions will be examined in the book *Smart Antennas* in this series; unfortunately, though, there are few published measurements yet on this topic, and much depends on such data.

Quadratic Expansion

The plots of spatial autocorrelation invite a quadratic approximation for small displacements, just like the quadratic expansions for **autocorrelation function in time** due to Doppler spread, and **autocorrelation function in frequency** due to delay spread. Like those expansions, we will find that it depends on the rms measure of the corresponding spread - in this case the angular dispersion. Looks as though we have a simple and useful result coming up. From the approximation (8.3.10), we have

$$R_g(x) = e^{j \cdot \beta \cdot x \cdot \sin(\Theta)} \cdot \int_{-\frac{\Omega}{2}}^{\frac{\Omega}{2}} P(\phi + \Theta) \cdot e^{j \cdot \beta \cdot x \cdot \cos(\Theta) \cdot \phi} \, d\phi \qquad (8.3.17)$$

We are interested in the broadside direction, the one from which we obtain diversity. Therefore we set $\Theta=0$. The second derivative, evaluated at $x=0$, is

$$\frac{d^2}{dx^2} R_g(0) = -\left(\frac{2 \cdot \pi}{\lambda}\right)^2 \cdot \int_{-\frac{\Omega}{2}}^{\frac{\Omega}{2}} P(\phi) \cdot \phi^2 \, d\phi = -\left(\frac{2 \cdot \pi}{\lambda}\right)^2 \cdot \sigma_g^2 \cdot \phi_{rms}^2 \qquad (8.3.18)$$

where ϕ_{rms} is the rms angle spread, and σ_g^2 normalizes the angular power density. This gives us an easy series expansion

$$R_g(x) = R_g(0) + \frac{1}{2} \cdot \frac{d^2}{dx^2} R_g(0) \cdot x^2 + O(x^3) \qquad (8.3.19)$$

and resulting approximation

$$R_g(x) = \sigma_g^2 \cdot \left[1 - \frac{1}{2} \cdot \left(2 \cdot \pi \cdot \phi_{rms} \cdot \frac{x}{\lambda} \right)^2 \right] \qquad (8.3.20)$$

Therefore, when you estimate the diversity antenna correlations, all you need to know is the rms angle spread - a simple and useful result. As a quick application of it, we note from (8.3.20) that the correlation distance must scale inversely with the rms angle spread.

9. SIMULATING FADING CHANNELS

Why is a section on simulation tacked onto a course in mobile channel characteristics? Two reasons: first, it's an important topic, and it had to go somewhere in the set of courses on Mobile and Personal Communications; and second, the unique problems of simulating fading channel effects are tied closely to the channel statistics - and that means this course.

However, if you are not particularly interested in knowing how to do these simulations, stop right now. **Read no further!** On the other hand, you might have to run some simulations in your thesis or on the job, and the material of this section could save you a lot of time - both development time and actual simulation run time. So you might just want to have a look...

9.1 Sampling and SNR

Over the years, I have seen graduate students and working engineers spend hour after hour trying to get the SNR calibration right. There are so many ways to be 3 dB or 6 dB off the true value. Including fading only makes it worse. And accounting for sampling rate - the number of samples per symbol - can cause strong people to crumble and seek new employment in life situations with less pressure.

Well, perhaps I exaggerate. But you may find this section useful, even if it is quite basic.

9.2 Complex gain generation

A key element of fading channel simulations is an efficient way to generate a complex gain process with the right statistics. This section shows you the two most common ways to do it. One of them even has Mathcad functions that you can use in your own simulation work.

9.3 Importance sampling

The main problem with conventional Monte Carlo simulations of fading channels is that the only events of interest - like bit errors - take place during fades. That means that 99% or more of your simulation run is spent generating nothing useful. And that in turn means very long simulation runs. You start having to tie up every free computer on your network, and let them run over weekends, just to generate one more point on a curve. At some point, you realize that there are many parameters to vary, and consequently, many curves - and you suddenly recognize that it may take longer than you thought to graduate or to meet the next project milestone.

There is a way out. Importance sampling (IS) can reduce your simulation run times by one, and sometimes two, orders of magnitude. This section gives a brief outline of the method. It also provides Mathcad functions for IS complex gain generation, the central part of the technique.

9.3A Effect of β on Variance in Importance Sampling

This appendix obtains the variance of an IS simulation that uses our biasing method when the instantaneous BER depends exponentially on the instantaneous SNR, a common situation.

9.1 Sampling and SNR

In this section, we'll get three simple, but useful, results:

* signal and noise scaling with complex envelopes;
* SNR scaling in both flat and frequency-selective fading;
* the effect of discrete-time operation on SNR.

Also, we see them put to use in a small simulation written in Mathcad. It's at the end of this section. Incidentally, if you want to learn more about simulation, read [**Jeru92**].

Scaling the Complex Gain

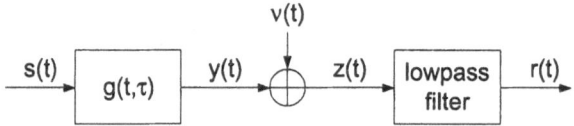

The sketch represents a typical link to be simulated. Most variations among systems at the link level are associated with the particular transmitted signal or the receiver processing, and are therefore outside the present discussion. Assume that the simulation uses complex envelopes to represent bandpass signals. The transmitted signal then has power

$$P_s = \frac{1}{2} \cdot E\left[(|s(t)|)^2 \right] = E_b \cdot R_b = E_s \cdot R_s \qquad (9.1.1)$$

where E_b and R_b are the energy per bit and the bit rate, and E_s and R_s are their per-symbol counterparts. The important part here is the factor 1/2, which relates the squared magnitude of the complex envelope to the power of the real bandpass signal it represents.

The received signal at the output of the fading channel, and before addition of receiver noise, is

for flat fading: for frequency selective fading:

$$y(t) = g(t) \cdot s(t) \qquad\qquad y(t) = \int_{-\infty}^{\infty} g(t,\tau) \cdot s(t-\tau) \, d\tau \qquad (9.1.2)$$

The power of the received signal is computed easily in the case of flat fading:

$$P_y = \frac{1}{2} \cdot E\left[(|y(t)|)^2 \right] = \frac{1}{2} \cdot E\left[(|g(t)|)^2 \right] \cdot E\left[(|s(t)|)^2 \right]$$

$$= 2 \cdot \sigma_g^2 \cdot P_s \qquad (9.1.3)$$

The point here is that the factor 1/2 can apply to only one of $s(t)$ or $g(t)$, so we have a factor of 2 to account for. That's the first pitfall, the one in which your SNR is 3 dB too high. My recommendation?

$\Longrightarrow$ Make the variance $\sigma_g^2 = \frac{1}{2}$; equivalently, make $E\left[\left(|\,g(t)\,|\right)^2\right] = 1$

so that the complex gain leaves the signal power unchanged.

In the case of frequency selective fading, the power of the channel output is

$$P_y = \frac{1}{2} \cdot E\left[\left(|\,y(t)\,|\right)^2\right] = \frac{1}{2} \cdot E\left(\int_{-\infty}^{\infty}\int_{-\infty}^{\infty} g(t,\tau)\cdot s(t-\tau)\cdot \overline{g(t,\alpha)\cdot s(t-\alpha)}\, d\tau\, d\alpha\right)$$

$$= \frac{1}{2} \cdot \int_{-\infty}^{\infty}\int_{-\infty}^{\infty} E\left(g(t,\tau)\cdot\overline{g(t,\alpha)}\right)\cdot E\left(s(t-\tau)\cdot\overline{s(t-\alpha)}\right) d\tau\, d\alpha$$

$$\tag{9.1.4}$$

In the case of **WSSUS channels**, we have

$$\frac{1}{2}\cdot E\left(g(t,\tau)\cdot\overline{g(t,\alpha)}\right) = P_g(\tau)\cdot\delta(\tau-\alpha) \tag{9.1.5}$$

so that

$$P_y = 2\cdot P_s \cdot \int_{-\infty}^{\infty} P_g(\tau)\,d\tau = 2\cdot\sigma_g^2\cdot P_s \tag{9.1.6}$$

This is the same as (9.1.3), and my recommendation is the same:

$\Longrightarrow$ Make the variance $\sigma_g^2 = \frac{1}{2}$; equivalently, make $E\left[\left(|\,g(t)\,|\right)^2\right] = 1$

In particular, if you are using a **TDL channel model** in your simulation - and you probably are - then you have

$$y(t) = \sum_{i=1}^{N} g_i(t)\cdot s(t-i\cdot\Delta t) \tag{9.1.10}$$

Following the same reasoning that led to (9.1.6), you ensure that the variances

$$\sigma_g^2 = \sum_{i=1}^{N} \sigma_i^2 \qquad \text{where} \qquad \sigma_i^2 = \frac{1}{2}\cdot E\left[\left(|\,g_i(t)\,|\right)^2\right] \tag{9.1.11}$$

leave the signal power unchanged. Simply:

$\Longrightarrow$ Make the variance $\sigma_g^2 = \frac{1}{2}$; that is, $\displaystyle\sum_{i=1}^{N}\sigma_i^2 = \frac{1}{2}$ $\tag{9.1.12}$

Sampling and Noise Scaling

Now we approach the other headache - scaling the sampled noise. First, look at the continuous time link model in the sketch above. The complex additive white Gaussian noise (AWGN) is $v(t)$, which has the property that its power spectrum is flat at N_o and its autocorrelation function is

$$R_v(\tau) = \frac{1}{2} \cdot E\left(\overline{v(t) \cdot v(t - \tau)}\right) = N_o \cdot \delta(\tau) \tag{9.1.13}$$

Usually, your fundamental SNR parameter is $\gamma_b = E_b/N_o$ or $\Gamma_b = E[\gamma_b]$. It doesn't affect your results in a linear system if the signal and noise are scaled by the same quantity, so that you can use (9.1.1) to calculate E_s and consequently N_o. Note that we assume that $\sigma_g^2 = \frac{1}{2}$; otherwise, you must multiply E_s by $2\sigma_g^2$.

Now convert this conceptual continuous time model to the discrete time model in your simulation. As shown in the sketch, we can imagine that the composit of received signal plus AWGN is filtered and then sampled. This notional filter has a rectangular frequency response with unit gain, and a passband $[-W, W]$ large enough to accommodate the signal without distortion. We sample at the Nyquist rate $2W$, giving independent identically distributed Gaussian noise samples $n(k)$, with variance

$$\sigma_n^2 = \frac{1}{2} \cdot E\left[(|n(k)|)^2\right] = 2 \cdot N_o \cdot W \tag{9.1.14}$$

Consequently, the samples $r(k)$ that you must generate in your simulation are just

$$r(k) = y\left(\frac{k}{2 \cdot W}\right) + n(k) \tag{9.1.15}$$

Usually, we choose W so that the sampling rate $2W$ is an integer multiple of the symbol rate.

And that's all there is to creating a discrete time simulation with the right SNR scaling. In summary:

1. On paper, determine an equation for $s(t)$, and the corresponding E_b or E_s.

2. From Γ_b or Γ_s, determine the corresponding value for N_o.

3. Choose a value for W that will accommodate the spread signal. That gives you the sample spacing $\Delta t = 1/2W$.

4. Generate the sampled signal $s(k\Delta t)$ directly from your equation in step 1.

5. Put the sampled signal through the TDL fading channel (9.1.10), in which the tap variances satisfy (9.1.12).

6. Add samples of white noise of variance σ_n^2 given by (9.1.14).

An Example

To make these suggestions more concrete, we'll work through a simple example: DPSK with rectangular pulses, differentially detected. The original data bits a(k) are ±1, and they are differentially encoded to form the transmitted amplitudes

$$c(k) = a(k) \cdot c(k-1) \tag{9.1.16}$$

which are also ±1. The transmitted signal is

$$s(t) = \sqrt{2 \cdot E_b} \cdot \sum_k c(k) \cdot p(t - k \cdot T) \tag{9.1.17}$$

where the pulse is rectangular with unit energy:

$$p(t) = \frac{1}{\sqrt{T}} \qquad \text{for} \qquad 0 \le t < T \qquad \text{and zero elsewhere.} \tag{9.1.18}$$

You can easily verify that E_b is the energy per bit. The continuous time receiver consists of a filter matched to the transmitted pulse $p(t)$, with its output sampled at the symbol rate to give the sequence $r(i)$. Subsequent differential detection recovers the estimate $a_{hat}(i)$ of the data bit $a(i)$:

$$a_{hat}(i) = if\left(Re\left(r(i) \cdot \overline{r(i-1)} \right) > 0, 1, -1 \right) \tag{9.1.19}$$

Now for the simulation. First, we pick an arbitrary value for P_s. This makes $E_b = P_s T$, and from the SNR, we have

$$N_0 = \frac{E_b}{\Gamma_b} = \frac{P_s \cdot T}{\Gamma_b} \tag{9.1.20}$$

As for the sampling rate, we can assume that the fictitious front end filter has bandwidth $W = 2/T$. At twice the symbol rate, it does little damage to the signal. The sampling rate $2W$ is therefore $4/T$; that is, we operate at

$$N_{ss} := 4 \qquad \text{samples per symbol} \tag{9.1.21}$$

and the signal sample i at the filter output for symbol k is

$$r(i) = \sqrt{2 \cdot E_b} \cdot c(k) \cdot \frac{1}{\sqrt{T}} \cdot g(i) = \sqrt{2 \cdot P_s} \cdot c(k) \cdot g(i) \tag{9.1.22}$$

From the above, the variance of the noise samples is

$$\sigma_n^2 = 2 \cdot W \cdot N_0 = \frac{N_{ss}}{T} \cdot \frac{P_s \cdot T}{\Gamma_b} = N_{ss} \cdot \frac{P_s}{\Gamma_b} \tag{9.1.23}$$

We can choose P_s to be any value we like, since the system is linear and the noise is scaled appropriately, leaving the BER unaffected. A convenient value is $1/T$, so that $E_b=1$, but that's not necessary - an equally convenient value, for other reasons, is $P_s=1$.

The receiver processing is simple: mimic the matched filter by adding N_{ss} successive received samples $r(i)$ and differentially detect the symbol spaced sums.

And finally, here's the actual simulation. First, some useful functions:

BPSK data $a(k)$ generator:

$$data2(k) := if(rnd(1)>0.5, 1, -1)$$

unit variance complex white noise generator - see **Appendix H** for a brief discussion:

$$cgauss(x) := \sqrt{-2 \cdot \ln(rnd(1))} \cdot \exp(j \cdot rnd(2 \cdot \pi))$$

variance 1/2 complex gain generator - see **Appendix B** for full details:

⬡ Reference:D:\COURSES\MobChann\paperbook\Jakesgen.mcd(R)

decision device:

$$slice(x) := if(x>0, 1, -1)$$

Choose the overall parameters:

arbitrary power: $P_s := 1$

samples per symbol: $N_{ss} := 4$

number of symbols in the run: $N_{sim} := 1000$

The simulation itself is shown on the facing page. Note that signal and noise scaling A and σ_n, respectively, follow the rules. The main symbol loop is simple: generate a differentially encoded bit c, generate a matched filter output MF, differentially detect and count any resulting error. The initialization simply sets the first differentially encoded bit and the first matched filter output.

$$\text{BERsim}\left(\Gamma_b, \text{fDT}\right) :=$$
$$A \leftarrow \sqrt{2 \cdot P_s}$$

$$\sigma_n \leftarrow \sqrt{N_{ss} \cdot \frac{P_s}{\Gamma_b}}$$

$$G \leftarrow \text{Jakes\_init}(1)$$

$$c_{last} \leftarrow -1$$

$$\eta \leftarrow \frac{\text{fDT}}{N_{ss}}$$

$$\text{errors} \leftarrow 0$$

$$MF_{last} \leftarrow \left[\sum_{i=0}^{N_{ss}-1} A \cdot c_{last} \cdot \text{Jakes\_gen}(\eta \cdot i, G) + \sigma_n \cdot \text{cgauss}(i) \right]$$

$$\text{for } k \in 1 .. N_{sim}$$

$$\quad a \leftarrow \text{data2}(k)$$

$$\quad c \leftarrow a \cdot c_{last}$$

$$\quad MF \leftarrow \left[\sum_{i=N_{ss} \cdot k}^{N_{ss} \cdot (k+1) - 1} A \cdot c \cdot \text{Jakes\_gen}(\eta \cdot i, G) + \sigma_n \cdot \text{cgauss}(i) \right]$$

$$\quad a_{hat} \leftarrow \text{slice}\left(\text{Re}\left(MF \cdot \overline{MF_{last}} \right) \right)$$

$$\quad (\text{errors} \leftarrow \text{errors} + 1) \text{ if } a_{hat} \neq a$$

$$\quad MF_{last} \leftarrow MF$$

$$\quad c_{last} \leftarrow c$$

$$\frac{\text{errors}}{N_{sim}}$$

For reference, the theoretical BER, from the text *Detection and Diversity* in this series, is

$$\text{BERtheor}\left(\Gamma_b, \text{fDT}\right) := \frac{1}{2} \cdot \frac{1 + \Gamma_b \cdot (1 - J0(2 \cdot \pi \cdot \text{fDT}))}{1 + \Gamma_b}$$

Let's try a few runs and compare them. To perform another run, put the cursor on BERsim below and press F9.

$\Gamma_b := 10$ $fDT := 0.01$

$\text{BERsim}(\Gamma_b, fDT) = 0.031$ $\text{BERtheor}(\Gamma_b, fDT) = 0.046$

Even for low SNR and high fade rate, where we generate a comparatively large number of errors, the agreement is pretty rough. We need much more than $N_{sim} = 1000$ for accuracy. At higher SNR, the error rate is smaller and we need even longer runs in order to get enough errors for accurate BER estimation. As for fade rate, the effect is more subtle. You know from **Section 4.3** that errors occur in bursts, during deep fades. Therefore it is the number of fades, more than the number of bits, that determines the accuracy. For very low fade rates, there may be only one fade, or even no fades, in a run. Unless we substantially increase the length of the run, accuracy will be very poor.

The problem of long simulation runs in fading channel simulations is solved in **Section 9.3**, where we deal with importance sampling.

9.2 Complex gain generation

As a central part of your channel simulation, you need a reliable way to generate channel complex gain samples with the right properties. Those properties are:

* The real and imaginary components are both Gaussian with the same autocorrelation function, but are independent of each other. If you want to represent frequency offset, which introduces correlation of the components, do it explicitly with a separate operation.

* Different complex gain generators (e.g., for different TDL taps in a **WSSUS frequency selective channel model**) should be independent.

* The autocorrelation function of the samples should be proportional to $J_0(2\pi f_D \Delta tk)$, as in **Section 5.1**.

Let's assess that third point. Why go to such pains to simulate isotropic scattering, when we know that very few channels are truly isotropic? The answers are (1) that isotropic scattering removes one parameter, direction of travel, from the general clutter of parameters; and (2) that it is almost universally used as a benchmark by which researchers and designers can compare the performance of different techniques.

We'll look at two widely used methods of complex gain generation: the Jakes method and the filtered noise method.

The Jakes Method

The Jakes method is simple, it has a low computational load, and it can be run forward or backward, in equispaced or irregular spacing. The spectrum of its complex gain is almost exactly that of the ideal spectrum and its pdf, for a dozen or more components in the sum, is generally considered adequately Gaussian. The method is also interesting because it mimics the very **interfering plane wave** situation that is at the heart of channel fading. You will find a detailed description in **Appendix B**, as well as the two functions you need in order to to use it in your own simulations.

An occasional criticism of the Jakes method is that it has difficulty ensuring very low correlation between different generators, because it depends on the luck of the draw in selecting initial phases. However, the **simple modification in Appendix B** solves this problem - essentially, by ensuring that the sets of arrival angles of the simulated plane waves are disjoint for different generators. The result is that different generators are uncorrelated, regardless of the choice of initial phases.

Here's how you use the two functions of **Appendix B** (you saw them before in the simulation of **Section 9.1** and at many other points in this text).

▣ Reference:D:\COURSES\MobChann\paperbook\Jakesgen.mcd(R)

$G := \text{Jakes\_init}(1)$ $i := 0.. 100$ $g_i := \text{Jakes\_gen}(0.1 \cdot i, G)$

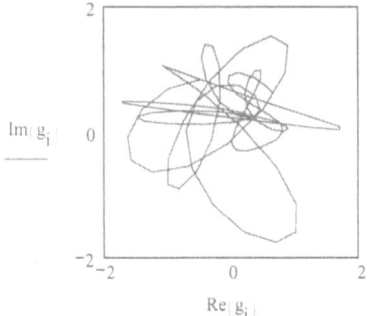

This is a typical sample function. **Note that Jakes_gen produces samples with variance 1/2,** to preserve signal power. To see different sample functions, place the cursor on the *G* initialization and recompute by pressing F9.

Filtered White Noise

The other popular method for complex generation is sketched below. Essentially, two Gaussian white noise generators pass through identical spectrum shaping filters to create the real and imaginary components at a relatively low sampling rate. The filters are followed by interpolators to bring the sampling rate up to the value required by the signal representations. You'll find efficient FFT-based procedures for generation of the gain process, wth full discussion, in **Appendix K**.

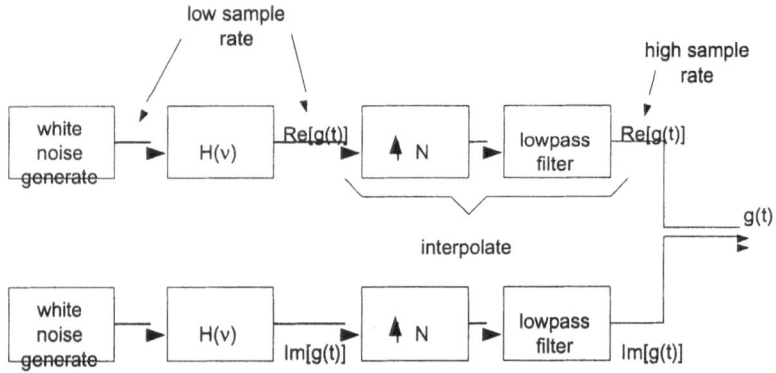

The filter transfer function is the square root of the desired spectrum shape. From **Section 5.1**, we have

$$H(v) = \sqrt{S_g(v)} = \begin{vmatrix} \dfrac{\sigma_g}{\sqrt{\pi \cdot f_D}} \cdot \dfrac{1}{\left[1 - \left(\dfrac{v}{f_D} \right)^2 \right]^{0.25}} & \text{if } |v| < f_D \\ 0 & \text{otherwise} \end{vmatrix} \qquad (9.2.1)$$

Eq. (9.2.1) exposes the weakness of the method. We have a bandlimited filter with strong singularities at the band edges. Approximating it closely requires a very high order FIR or IIR filter (many tens, or even hundreds, of taps). If implemented without thought, it poses a significant computational load that is often comparable to the load of the receiver processing algorithms to be tested (however, see **Appendix K** for an efficient variant). In addition, it is not flexible: input and output samples must be equispaced, and it requires tricks to go backward, if that is desired.

Because of the computational load, the spectrum shaping filter is sometimes cascaded with an interpolator. In this way, the basic samples of complex gain can be generated at a low sampling rate - a few multiples of the Doppler frequency - with correspondingly shorter filters. A comparatively simple minimum mean squared error (MMSE) interpolator takes it up to the signal sampling rate, which is usually about 4 times the symbol rate.

A point in its favour is that the output samples in the filtered noise method are Gaussian; in fact, they are, if anything, even more Gaussian than the input samples, because of the central limit theorem effect of the filters. Further, the generated process is stationary. Both of these propeties are in contrast to the Jakes method. In addition, there is no difficulty ensuring independence of different generators, simply by use of different white noise sources; this property is also in contrast to published versions of Jakes' method (although not to the variant provided in **Appendix B**).

9.3 Importance sampling

The main problem with conventional Monte Carlo simulations of fading channels is that the only events of interest - such as bit errors - take place during fades, as we saw in **Section 4.3**. However, deep fades are rare; for example, the SNR is 20 dB or more below its mean value only 1% of the time, a fact you learned in **Section 4.2**, and 30 dB below only 0.1% of the time. If significant numbers of errors show up only in fades that are 20 dB or deeper, 99% or more of your simulation run is spent uselessly generating bits that, by comparison, almost never experience an error. Effectively, it is the number of fades more than the number of bits, that determines accuracy. And that in turn means very long simulation runs, especially when the fade rate is very low.

Fortunately, there is a way to shorten simulation runs dramatically. Importance sampling (IS) can reduce your simulation run times by one, and sometimes two, orders of magnitude. This section gives:

* a brief outline of an IS method for fading channels, presented originally in [**Cave92**];
* a description of the Mathcad functions for IS complex gain generation, the central part of the technique, which are are contained in **Appendix H** ;
* an example of IS applied to the same DPSK system simulated in Section 9.1.

For a more general discussion of IS in other contexts, see [**Jeru92, Smit97, AlQa97, Leta97**].

The Important Parts of Importance Sampling

In general, IS becomes attractive whenever rare events produce most of the activity you are recording, and therefore dominate the sample averages. That is certainly the case here, since deep fades are rare events, but produce bursts of errors. The problem? Too much scatter, or variability, in the simulation results. The remedy? Alter, or "bias", the complex gain statistics so that deep fades are produced more often than normal. That gives more errors and greater statistical stability for a given run length. Of course, we also need a method to transform this artificially high error rate back to more normal values, using the known bias statistics.

To use IS, we must organize the simulation as a set of short runs, each with independently selected random variables, and average the BER over the set of runs. By itself, this is not an improvement. In fact, it worsens matters somewhat, since many receivers have memory or other end effects. Excising these transients reduces the effective number of bits, which affects short runs more than a single long run.

To begin, recognize that the simulation is a deterministic function of the set of random variables. In a typical case, we have three ensembles: the data, fading and noise sequences, denoted c, g, and n, respectively. Define the random vector

$$x = (c \quad g \quad n)$$

(9.3.1)

with pdf

$$p_x(x) = p_c(c) \cdot p_g(g) \cdot p_n(n)$$

(9.3.2)

Then associated with each x is a decoded data vector $c_{hat}(x)$ and a sample BER $h(x)$ that depends on c and c_{hat}. This sample BER is a deterministic function of x. The true BER is the mean

$$e = E_x(h(x)) = \int h(x) \cdot p_x(x) \, dx \tag{9.3.3}$$

where $E_x[\ ]$ denotes expectation with respect to the pdf $p_x(x)$.

First, consider an ordinary unbiased simulation consisting of N short runs with random vectors x_i, $i=1..N$. The estimated BER is

$$e_{hat} = \frac{1}{N} \cdot \sum_{i=1}^{N} h(x_i) \tag{9.3.4}$$

This estimator is unbiased, since its mean is the true BER:

$$E_x(e_{hat}) = e \tag{9.3.5}$$

Its variance is therefore

$$\sigma^2 = E_x\left[(e_{hat} - e)^2\right] = E_x\left[(e_{hat} - e) \cdot e_{hat}\right] \qquad \text{since it's unbiased}$$

$$= E_x\left[\left(\frac{1}{N} \cdot \sum_{i=1}^{N} h(x_i) - e\right) \frac{1}{N} \cdot \sum_{i=1}^{N} h(x_i)\right]$$

$$= E_x\left[\frac{1}{N} \cdot (h(x) - e) \cdot h(x)\right] \qquad \text{since runs are independent}$$

$$= \frac{1}{N} \cdot \int (h(x) - e) \cdot h(x) \cdot p_x(x) \, dx \tag{9.3.6}$$

Next, we turn to simulation with a biased pdf. That simply means that we generate the random vectors x_i with a different pdf

$$p_{x\_b}(x) = \frac{p_x(x)}{w(x)} \tag{9.3.7}$$

where the weight $w(x)$ can be chosen to emphasize selected regions of the x space. In our case, we would make $w(x)$ small in the fade regions that generate most of the errors, in order to increase their probability of being generated. The action of our simulated link follows the same deterministic function $h(x)$ as before, and if we simply average the sample BERs, the result will have the wrong mean (too large, if we emphasize error-producing regions) - that is, it will be biased.

Now here's the trick. We will weight, or de-emphasize, the sample BER $h(x)$ by $w(x)$ before including it in the average. This transforms the biased BERs back to the true statistics. We'll allow ourselves a possibly different number of runs N_b, and calculate the sample mean value as

$$e\_{hat\_b} = \frac{1}{N_b} \cdot \sum_{i=1}^{N_b} w(x_i) \cdot h(x_i) \tag{9.3.8}$$

Because we are averaging with random vectors generated by the biased pdf, this BER estimate is unbiased:

$$E_{x\_b}(e\_{hat\_b}) = E_{x\_b}(w(x) \cdot h(x)) \qquad \text{since runs are independent}$$

$$= \int w(x) \cdot h(x) \cdot p_{x\_b}(x)\, dx \qquad \text{from (9.3.7)}$$

$$= \int h(x) \cdot p_x(x)\, dx = e \tag{9.3.9}$$

The next question is whether the sample variance is affected - that is, whether IS is worthwhile. Following the same logic as in (9.3.6), we have the variance from biased generation as

$$\sigma_b^2 = \frac{1}{N_b} \cdot \int (w(x) \cdot h(x) - e) \cdot w(x) \cdot h(x) \cdot p_{x\_b}(x)\, dx$$

$$= \frac{1}{N_b} \cdot \int (w(x) \cdot h(x) - e) \cdot h(x) \cdot p_x(x)\, dx \tag{9.3.10}$$

We can compare this directly with the estimation error variance (9.3.6) in the unbiased simulation. If $w(x)=1$, then they have the same variance. However, if we choose $w(x) \approx e/h(x)$ then the variance in the biased simulation is dramatically reduced for the same number of runs. As expected, this weighting emphasizes regions in which $h(x)$ is large. If we could achieve equality, $w(x)=e/h(x)$, then the variance would be zero - but, unfortunately, that would require knowledge of e, the quantity we built the simulation to estimate!

We can therefore take advantage of IS to reduce the estimation error in the BER statistics or the number of runs or both. It was applied to multi-symbol differential detection in [**Cave92**], where a saving of one to two orders of magnitude in number of runs was achieved.

How to Bias Complex Gain Generation

To apply IS to fading channel simulations, we need a way to produce more fades in our generation of the complex gain sequences g. We'll do it with a biasing strategy that puts the centre sample of the gain sequence roughly in the middle of a fade. Define

$$G = \begin{bmatrix} g_{-K} & \bullet\ \bullet & g_{-1} & g_1 & \bullet\ \bullet & g_K \end{bmatrix} \tag{9.3.11}$$

as the $2K$-tuple obtained by deleting the centre sample g_0 from g. Also denote the pdf of g_0 and the conditional probability of G as

$$p_0(g_0) \quad \text{and} \quad p_G(G \mid g_0) \tag{9.3.12}$$

Clearly, the pdf of g is the product

$$p_g(g) = p_G(G \mid g_0) \cdot p_0(g_0) \tag{9.3.13}$$

To put a fade in the middle of g, we generate g_0 with a biased density $p_{0\_b}(g_0)$ that generates smaller values. The biased pdf of the vector g is therefore

$$p_{g\_b}(g) = p_G(G \mid g_0) \cdot p_{0\_b}(g_0) \tag{9.3.14}$$

One easy way to bias the density is to make its variance smaller by a factor β. Therefore

$$p_{0\_b}(g_0) = \frac{1}{2 \cdot \pi \cdot \beta \cdot \sigma_g^2} \cdot \exp\left[-\frac{|g_0|^2}{2 \cdot \beta \cdot \sigma_g^2} \right] \tag{9.3.15}$$

and the weight function is

$$w(x) = \beta \cdot \exp\left[\frac{(1 - \beta) \cdot \left(|g_0|\right)^2}{2 \cdot \beta \cdot \sigma_g^2} \right] \tag{9.3.16}$$

Having selected the centre value according to the biased pdf (9.3.15), all that's left is to generate the rest G of the complex gain vector g according to the conditional pdf. The idea is simple: "grow" the rest of the gain vector by adding a sample to the left of the midpoint, then a sample to the right of the midpoint. Now we have three samples. Add a sample to the left of this trio, then to the right, giving five. And so on. How do we ensure the right statistics? The mean value of each sample is obtained by linear estimation from the previously generated samples, and we add to it a sample of random noise having the right estimation error variance. The method is described more fully in [**Cave92**] and **Appendix H**, which contains the required functions.

It is important to recognize that this method is not simply equivalent to a change in SNR; because phase hits are associated with deep fades, it also gives greatly improved accuracy in the error floor region of the BER curves.

An Example

To demonstrate IS, we'll use the same binary DPSK example that we used at the end of **Section 9.1**. Recall that the original data bits $a(k)$ are ± 1, and they are differentially encoded to form the transmitted amplitudes $c(k)$. We operate at N_{ss} samples per symbol and the signal sample i at the filter output for symbol k is

$$r(i) = \sqrt{2 \cdot E_b} \cdot c(k) \cdot \frac{1}{\sqrt{T}} \cdot g(i) = \sqrt{2 \cdot P_s} \cdot c(k) \cdot g(i) \tag{9.3.17}$$

The variance of the noise samples is

$$\sigma_n^2 = 2 \cdot W \cdot N_o = \frac{N_{ss}}{T} \cdot \frac{P_s \cdot T}{\Gamma_b} = N_{ss} \cdot \frac{P_s}{\Gamma_b} \tag{9.3.18}$$

and we can choose P_s to be any value we like - such as 1. The receiver processing is simple: mimic the matched filter by adding N_{ss} successive received samples $r(i)$ and differentially detect the symbol spaced sums.

Our runs will be only two bits long. Why? First, simplicity in the demonstration. Second, it highlights the loss due to end effects, since the two transmitted bits carry only one data bit. In a real situation, you would make the runs a little longer, to make it more efficient.

And here's the actual simulation. First, some useful functions:

BPSK data $a(k)$ generator:

```
data2(k) := if(rnd(k)>0.5,1,-1)
```

unit variance complex white noise generator - see **Appendix H** for a brief discussion:

```
cgauss(x) := √(-2·ln(rnd(1)))·exp(j ·rnd(2·π))
```

variance 1/2 complex gain generator:

▣ Reference:D:\COURSES\MobChann\paperbook\Isgen.mcd(R)

which makes $\sigma_g := \dfrac{1}{\sqrt{2}}$

decision device:

```
slice(x) := if(x>0,1,-1)
```

From (9.3.16), the weight function is

$$w(g,\beta) := \beta \cdot \exp\left[\frac{(1-\beta) \cdot (|g|)^2}{2 \cdot \beta \cdot \sigma_g^2}\right] \tag{9.3.20}$$

Here on one page is the full simulation:

$$\text{BERsim}\left(\Gamma_b, \text{fDT}, N_{\text{sim}}, \beta\right) := \begin{array}{|l}
A \leftarrow \sqrt{2 \cdot P_s} \\[2mm]
\sigma_n \leftarrow \sqrt{N_{ss} \cdot \dfrac{P_s}{\Gamma_b}} \\[3mm]
\eta \leftarrow \dfrac{\text{fDT}}{N_{ss}} \\[3mm]
\sigma_0 \leftarrow \sqrt{\beta} \cdot \sigma_g \\[2mm]
\text{for } i \in 0..N-1 \\[1mm]
\quad R_i \leftarrow \dfrac{1}{2} \cdot J0(2 \cdot \pi \cdot \eta \cdot i) \\[2mm]
U \leftarrow \text{levinson}(R) \\[1mm]
\text{BERsum} \leftarrow 0 \\[1mm]
c_{\text{old}} \leftarrow -1 \\[1mm]
\text{for } k \in 1..N_{\text{sim}} \\[1mm]
\quad \begin{array}{|l}
a \leftarrow \text{data2}(k) \\[1mm]
c \leftarrow a \cdot c_{\text{old}} \\[1mm]
g \leftarrow \text{IS\_gen}\left(U, \sigma_0\right) \\[1mm]
\text{MF}_{\text{old}} \leftarrow \left[\displaystyle\sum_{i=0}^{N_{ss}-1} A \cdot c_{\text{old}} \cdot g_i + \sigma_n \cdot \text{cgauss}(i) \right] \\[4mm]
\text{MF} \leftarrow \left[\displaystyle\sum_{i=N_{ss}}^{2 \cdot N_{ss}-1} A \cdot c \cdot g_i + \sigma_n \cdot \text{cgauss}(i) \right] \\[4mm]
a_{\text{hat}} \leftarrow \text{slice}\left(\text{Re}\left(\text{MF} \cdot \overline{\text{MF}_{\text{old}}} \right) \right) \\[1mm]
\text{error} \leftarrow \text{if}\left(a_{\text{hat}} \neq a, 1, 0 \right) \\[1mm]
\text{weighted\_error} \leftarrow w\left(g_{N_{ss}}, \beta \right) \cdot \text{error} \\[1mm]
\text{BERsum} \leftarrow \text{BERsum} + \text{weighted\_error}
\end{array} \\[2mm]
\dfrac{\text{BERsum}}{N_{\text{sim}}}
\end{array}$$

The simulation itself is shown on the previous page. Most of the variables have the same interpretation as they did in the simulation of Section 9.1. However, σ_0 is the biased standard deviation of the centre sample of the gain vector **g**.

Choose the overall parameters:

arbitrary power: $P_s := 1$ number of short runs in simulation $N_{sim} := 1000$

samples per symbol: $N_{ss} := 4$ number of bits in each run $N_{bits} := 2$

We need $N_{bits} \cdot N_{ss} = 8$ gain samples in each run, but we'll have to generate one more, because the IS gain generation method makes it odd.

$$N := N_{bits} \cdot N_{ss} + 1$$

For reference, the theoretical BER, from the text *Detection and Diversity* in this series, is

$$BERtheor(\Gamma_b, fDT) := \frac{1}{2} \cdot \frac{1 + \Gamma_b \cdot (1 - J0(2 \cdot \pi \cdot fDT))}{1 + \Gamma_b}$$

Let's see if it works. You can change the parameters below and compare the simulated and true error rates. Note that $\beta = 1$ corresponds to unbiased simulation. To perform another run without changing parameters, put the cursor on BERsim below and press F9 - try it a few times!

$\Gamma_b := 100$ $fDT := 0.001$ $\beta := 1$ $N_{sim} := 1000$

$BERsim(\Gamma_b, fDT, N_{sim}, \beta) = 6 \cdot 10^{-3}$ $BERtheor(\Gamma_b, fDT) = 4.955 \cdot 10^{-3}$

This is fun, but we need a more systematic demonstration of the variance reduction properties of importance sampling. One easy way is to divide the total number of trials into, say, 10 subruns and look at the scatter. We can then compare unbiased ($\beta = 1$) and biased ($\beta < 1$) simulations.

$N_{runs} := 10$ $k := 0 .. N_{runs} - 1$

unbiased   **biased** $\beta := 0.1$

$$unbias_k := BERsim\left(\Gamma_b, fDT, \frac{N_{sim}}{N_{runs}}, 1\right)$$ $$bias_k := BERsim\left(\Gamma_b, fDT, \frac{N_{sim}}{N_{runs}}, \beta\right)$$

$mean(unbias) = 7 \cdot 10^{-3}$ Put the cursor on each $mean(bias) = 5.391 \cdot 10^{-3}$

of the two simulations

$stdev(unbias) = 6.403 \cdot 10^{-3}$ and press F9 a few $stdev(bias) = 1.556 \cdot 10^{-3}$

times.

It works! You should see that the biased simulation is closer to the theoretical value above and has less scatter than the usual unbiased simulation.

How Can We Choose a Good Value of β?

You have just seen that the variance of the BER estimate varies substantially with β. But is a smaller β value always better? How can you choose a good value? This section provides some answers: first, smaller isn't always better - there is an optimum value - and second, we can obtain some reasonable guidelines for selection.

We saw in (9.3.10) that there is a weight function $w(x)$ that results in an IS estimate of zero variance. In a more realistic situation, we select our weight function from a parameterized family, such as (9.3.16) and (9.3.20), which are determined by β. As the parameter shifts us from unbiased simulation, the increasing emphasis of the error producing region at low SNR decreases the variance, as expected; however, overemphasis of the region causes the variance to increase again. Note that the estimate e_{hat_b} (9.3.8) remains unbiased, regardless of the value of β.

Look over the **analysis of the effect of β on scatter** of our DPSK simulation, if you wish. Briefly, it shows that

* in the "underbiased" region, where β is greater than its optimum value, the variance is roughly proportional to β;
* in the "overbiased region", where β is smaller than its optimum value, the variance increases very rapidly, and may not even exist if β is sufficiently small.

Thus the penalty for overbiasing is far greater than for underbiasing, and it pays to be cautious.

Here's what happens if you overbias the importance sampling by choosing too small a value of β. The usual effect is that, over a small number of runs, the error rate is much too low. This is because deep fades are produced frequently, but the bias correction function $w(g)$ (9.3.20) savagely discounts them. On the other hand, over many, many, many runs, the average will be correct, since the estimate remains unbiased, as noted above. This is because fading peaks (the opposite of deep fades), which have become exceedingly rare events in the biased gain pdf, receive a huge weight $w(g)$. We are back to a "rare events producing large terms" scenario, with its attendant large variance, that prompted us to employ IS in the first place - and the scatter increases dramatically as β is reduced below its optimum value. Try it in the simulations above.

Assuming that we are not in the overbiased region, we can develop some guidelines for choice of β. First, consider how β must vary with the average SNR Γ in slow fading. We know that the error-producing region corresponds to deep fades - roughly, below some "trouble level" of instantaneous SNR. Recall from the discussion on **Rayleigh fading** that the probability of such an event decreases inversely with average signal strength, or average SNR. Therefore, to keep the estimation error variance more or less constant as SNR changes, we keep the deep fade event probability roughly constant, which implies that

===> β should be varied inversely with Γ for slow fading (i.e. in the noise limited region)

On the other hand, in the error floor region the BER does not decrease with increases in SNR. Consequently, continued decrease of β will overbias the simulation, causing the estimation error variance to increase again rapidly.

It is also possible to change β adaptively [**AlQa97**] over the course of many simulation runs in order to minimize the scatter.

Here is some simple advice for choosing β manually. At some modest level of Γ, say 10 dB, perform a few runs without biasing the pdf; that is, with $\beta=1$. Calculate the scatter. Now reduce β by degrees, performing a few runs at each value to estimate the mean and scatter. You should see the scatter drop. Below some value of β, the estimated mean will drop sharply, indicating that you have gone too far. Back off from that value, remembering that it is safer to underbias than to overbias. Having selected a good value of β for $\Gamma=10$ dB, you can follow the inverse scaling rule above as you increase Γ, until you hit the error floor.

So good luck! Hope this lets you graduate a little sooner.

9.3A Effect of β on Variance in Importance Sampling

This appendix obtains the variance of an IS simulation that uses our biasing method when the instantaneous BER depends exponentially on the instantaneous SNR; that is,

$$h(x)=a\cdot e^{-b\cdot x} \tag{9.3A.1}$$

where $x =|g|^2$, the squared magnitude of the instantaneous channel gain. This form of the instantaneous error rate, with parameters a and b, is a good model for many incoherent detection methods. For example, the DPSK detector in our simulation has a bit error rate of 0.5 exp$(-\Gamma x)$ in static channels, where Γ is the SNR when $x=1$. Even for coherent detection, where the BER usually has a Q function dependence on γ, appropriate choice of parameters a and b can give a good approximation.

The bias method we have adopted is simply to reduce the mean value of squared gain by a factor β, so that the biased pdf and the weight function are, respectively,

$$p_{x\_b}(x)=\frac{1}{\beta}\cdot\exp\left(\frac{-x}{\beta}\right) \qquad \text{and} \qquad w(x)=\beta\cdot e^{\left(\beta^{-1}-1\right)\cdot x} \tag{9.3A.2}$$

Note that the pdf has the correct unbiased form

$$p_x(x)=\exp(-x) \tag{9.3A.3}$$

when $\beta=1$, since we set $\sigma_g^2=1/2$ as a convention in Section 9.1.

To determine the variance of the simulation result, all we need are the first two moments of the weighted instantaneous BER $w(x)h(x)$. From (9.3.8) and (9.3.9), we have the first moment (the mean) as

$$\mu_1=\int_0^\infty w(x)\cdot h(x)\cdot p_{x\_b}(x)\,dx=\int_0^\infty h(x)\cdot p_x(x)\,dx=\int_0^\infty a\cdot e^{-b\cdot x}\cdot e^{-x}\,dx=\frac{a}{b+1}$$

$$\tag{9.3A.4}$$

whether the pdf is biased or not. The second moment of $w(x)h(x)$ needs a little more care; for a single run ($N_b=1$), it is

$$\mu_2=\int_0^\infty w(x)^2\cdot h(x)^2\cdot p_{x\_b}(x)\,dx=\int_0^\infty w(x)\cdot h(x)^2\cdot p_x(x)\,dx$$

$$I = \int_0^\infty \beta \cdot e^{\left(\beta^{-1} - 1\right) \cdot x} \cdot a^2 \cdot e^{-2 \cdot b \cdot x} \cdot e^{-x} \, dx = \beta \cdot a^2 \cdot \int_0^\infty \exp\left[-\left(2 \cdot b + 2 - \beta^{-1}\right) \cdot x\right] dx$$

$$I = \frac{\beta \cdot a^2}{2 \cdot b + 2 - \beta^{-1}} \qquad \text{for} \qquad \beta > \frac{1}{2 \cdot (b + 1)} \tag{9.3A.5}$$

Notice the restriction on β; if it is smaller than $2(b+1)$, the integral does not converge and the second moment and the variance are infinite. This is a very clear warning about overbiasing the pdf! Assuming that the condition is satisfied, the variance of a single biased run is the usual expression in terms of first and second moments (9.3A.4) and (9.3A.5), given by

$$\sigma_b^2 = \mu_2 - \mu_1^2 = \frac{\beta \cdot a^2}{2 \cdot b + 2 - \beta^{-1}} - \left(\frac{a}{b + 1}\right)^2 \tag{9.3A.6}$$

Let's see how the variance depends on β. Define it as a function

$$\text{var}_b(\beta, a, b) := \frac{\beta \cdot a^2}{2 \cdot b + 2 - \beta^{-1}} - \left(\frac{a}{b + 1}\right)^2 \tag{9.3A.7}$$

and input the average SNR in dB: $\Gamma_{dB} := 10$ $\Gamma := 10^{0.1 \cdot \Gamma_{dB}}$ (try other values, too)

For DPSK, the parameters of the instantaneous BER expression are

$$a := 0.5 \qquad b := \Gamma$$

Then $\beta_{min} := \dfrac{1}{2\,(b + 1)}$ from (9.3A.5), and we set up the plot parameters as

$$N_p := 500 \qquad \text{points in the plot} \qquad i := 0 .. N_p$$

$$\beta_i := \frac{1 - \beta_{min}}{N_p} \cdot i + \beta_{min} \cdot 1.0001$$

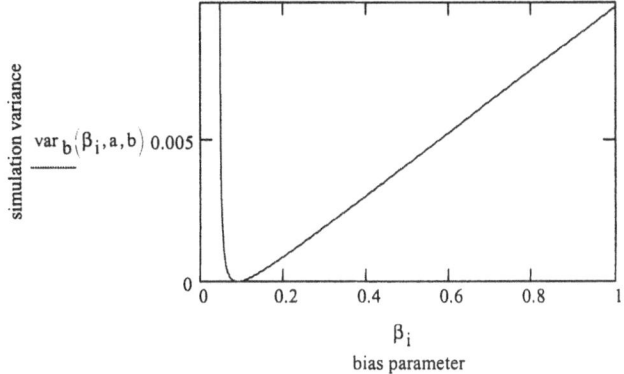

Very interesting - the variance is proportional to β when underbiased, and rises *very* steeply when overbiased. This plot illustrates both the value of importance sampling and the dangers of overbiasing. The minimum variance in this example is zero - yes, zero - because the biased pdf and the instantaneous BER both happen to be exponential. Such a good match is not common, unfortunately.

Usually we average many runs, so the resulting sample variance depends inversely on the number of runs N_b. The above graph of variance for a single run therefore implies that, for a given simulation accuracy, the required number of runs is approximately proportional to β in the underbiased region. For reference, we can normalize the curve by its unbiased value (i.e., β=1) to obtain a "run reduction factor"

$$R(\beta, a, b) := \frac{\mathrm{var}_b(\beta, a, b)}{\mathrm{var}_b(1, a, b)}$$

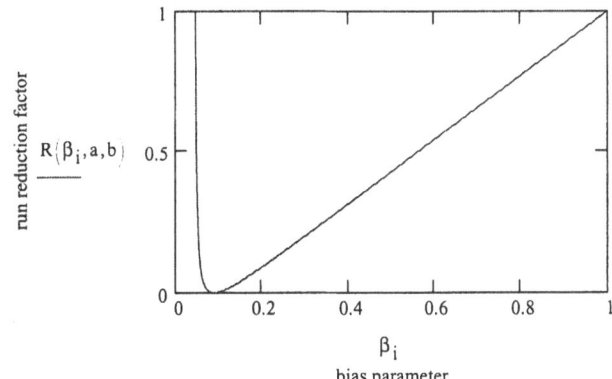

The variance plot is useful, of course, but another view of its significance is the *coefficient of variation*, defined as the ratio of standard deviation to mean

$$C = \frac{\sigma_b}{\mu_1} = \frac{b+1}{a} \cdot \sqrt{\frac{\beta \cdot a^2}{2 \cdot b + 2 - \beta^{-1}} - \left(\frac{a}{b+1}\right)^2} = \sqrt{\frac{\beta \cdot (b+1)^2}{2 \cdot b + 2 - \beta^{-1}} - 1} \qquad (9.3A.8)$$

This measure expresses the scatter in simulation results as a fraction of the average error rate, so it is a natural measure of the accuracy of the simulation. Let's see how it varies with the bias parameter β. Define it as a function:

$$C(\beta, a, b) := \sqrt{\frac{\beta \cdot (b+1)^2}{2 \cdot b + 2 - \beta^{-1}} - 1} \qquad (9.3A.9)$$

and define the average SNR in dB: $\quad \Gamma_{dB} := 10 \qquad \Gamma := 10^{0.1 \cdot \Gamma_{dB}}$ (try other values, too)

Then for DPSK and a single run ($N_b=1$), the coefficient of variation *normalized by its unbiased value* (i.e., at $\beta=1$) looks like this

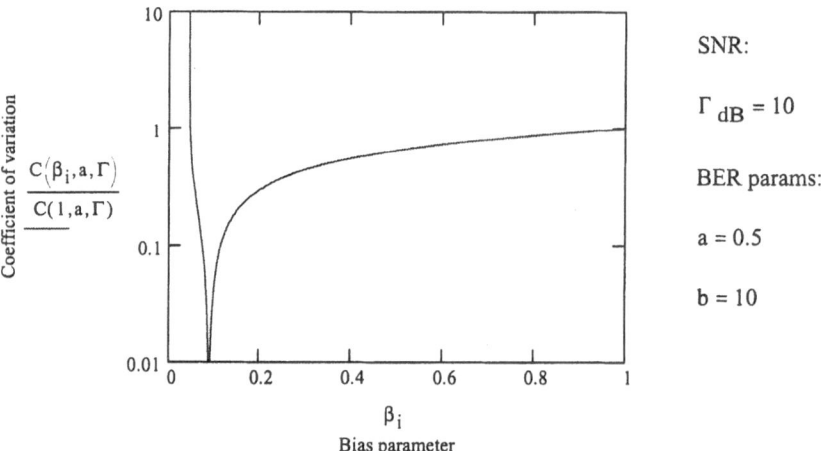

SNR:

$\Gamma_{dB} = 10$

BER params:

$a = 0.5$

$b = 10$

This illustrates the fact that significant improvement in accuracy is achieved only as β approaches the optimum point - and this is also a dangerous region, since imperfect setting can result in overbiasing.

APPENDIX A: THE LOGNORMAL DISTRIBUTION

Form of the density function

Although the lognormal distribution does not often appear in introductory texts on communications, it is very common in propagation studies (and many other fields). We say that a variable x is lognormally distributed if its logarithm y is normally (i.e., Gaussian) distributed. That is,

$$y = \ln(x) \qquad x = e^y \qquad p_y(y) = \frac{1}{\sqrt{2 \cdot \pi} \cdot \sigma_y} \cdot \exp\left[-\frac{1}{2} \left(\frac{y - \mu_y}{\sigma_y} \right)^2 \right] \tag{A1}$$

An elementary conversion gives the parameterized pdf of the original log-normal variable as

$$p_x(x, \mu_y, \sigma_y) := \frac{1}{\sqrt{2 \cdot \pi} \cdot \sigma_y \cdot x} \cdot \exp\left[-\frac{1}{2} \left(\frac{\ln(x) - \mu_y}{\sigma_y} \right)^2 \right] \quad \text{for } x \geq 0 \tag{A2}$$

Note that the mean $\mu_x \neq \exp(\mu_y)$, and the same with the standard deviation. Since decibels are a logarithmic measure, a lognormally distributed variate is Gaussian when expressed in dB:

$$z = 10 \cdot \log(x) \quad (z \text{ is } x \text{ in dB}) \qquad x = 10^{0.1 \cdot z} \qquad y = 0.1 \cdot \ln(10) \cdot z$$

$$p_z(z) = \frac{1}{\sqrt{2 \cdot \pi} \cdot \sigma_z} \cdot \exp\left[-\frac{1}{2} \left(\frac{z - \mu_z}{\sigma_z} \right)^2 \right] \qquad \text{where} \qquad \begin{aligned} \mu_z &= \frac{\mu_y}{0.1 \cdot \ln(10)} \\[2mm] \sigma_z &= \frac{\sigma_y}{0.1 \cdot \ln(10)} \end{aligned} \tag{A3}$$

Visualization

What does the pdf of a lognormal variate look like? Below, you can choose the mean and standard deviation of z, the dB equivalent of x, and see the effect on the graph.

$$\mu_z := 0 \text{ dB} \qquad \sigma_z := 6 \text{ dB} \qquad \text{(Note both these values are in dB, and } \mu_z \text{ can be negative)}$$

Convert to y for convenience:

$$\mu_y := 0.1 \cdot \ln(10) \cdot \mu_z \qquad \sigma_y := 0.1 \cdot \ln(10) \cdot \sigma_z$$

Other values for comparison: $\qquad \sigma_{y1} := 0.1 \cdot \ln(10) \cdot 1 \qquad \sigma_{y10} := 0.1 \cdot \ln(10) \cdot 10$

$$x := 0.01, 0.02 .. 4$$

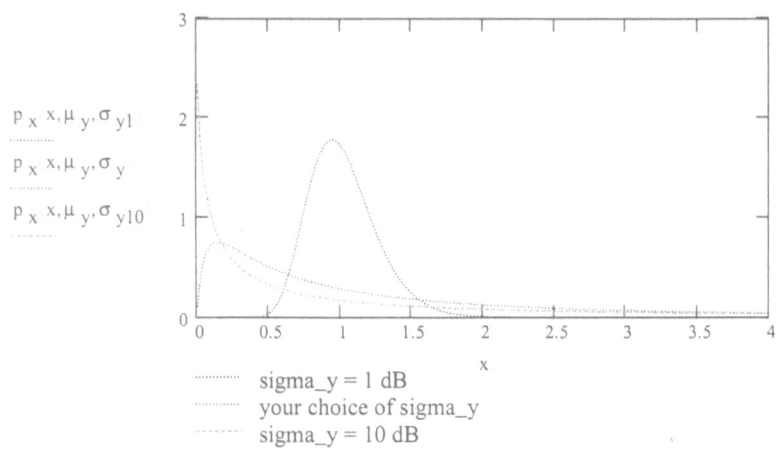

Log-Normal Probability Density Function

Try this: keep the mean μ_z at 0 dB and vary the standard deviation. You should see that for a standard deviation σ_z of 6 to 8 dB, typical for shadowing, the pdf is quite asymmetric, and with a small modal value. If you reduce the standard deviation, it becomes more symmetric, roughly centered on $x=1$. In fact, it tends to a Gaussian shape, since the range of x with significant probability becomes small enough that the logarithm in (A4) varies almost linearly in x.

Useful Properties

The lognormal distribution has some interesting properties:

* The product (or quotient) of lognormal variates is lognormal. That's because the exponents y or z add (or subtract). Since the exponents are normally distributed, they remain normal after addition (or subtraction).

* For this reason, both the power and the amplitude of a lognormal signal are lognormal.

* The product of a large number of independent, identically distributed variates is asymptotically lognormal, almost regardless of the distribution of the variates. This is the counterpart of the central limit theorem for addition of variates. The reason should be obvious at this point: the exponents add, so the central limit theorem applies to them, and their sum converges to a Gaussian distribution. Actually, independence and identical distribution are unnecessarily stringent requirements - see [**Papo84**] for a good discussion of the central limit theorem.

From these points, you can see the reason for the importance of this distribution in propagation studies. A typical signal has undergone several reflections or diffractions by the time it is received, each of which can be characterized by an attenuation, or multiplication. The cascaded multiplications converge to a lognormal distribution.

Moments

The moments of a lognormal variate are often useful in studies of signal detection in shadowed conditions. The nth moment of x is, by definition,

$$\mu_x^{(n)} = E_x(x^n) \tag{A.5}$$

where $E_x[\ ]$ denotes expectation over x. Substitution of (A.1) gives

$$\mu_x^{(n)} = E_y(e^{n \cdot y}) \tag{A.6}$$

Now recall that the characteristic function of a Gaussian random variable (the Fourier transform of its pdf) is well known, and given by

$$M_y(v) = E_y(e^{j \cdot v \cdot y}) = e^{j \cdot v \cdot \mu_y} \cdot e^{\frac{-\sigma_y^2 \cdot v^2}{2}} \tag{A.7}$$

Identify jv with n and you have the nth moment of x as

$$\mu_x^{(n)} = M_y(-j \cdot n) = \exp\left(n \cdot \mu_y + \frac{n^2 \cdot \sigma_y^2}{2}\right) \tag{A.8}$$

This is a very simple result. The exponential nature of the lognormal distribution has once again simplified a problem of products (powers of x, in this case).

Be Careful in Normalization

Finally, a potential pitfall associated with the lognormal pdf. Suppose a received signal power is lognormally distributed as

$$P = P_0 \cdot x \tag{A.9}$$

For convenience in analysis, you would normalize x to have unit mean. However, this does not imply that the exponent y has zero mean! Equivalently, the mean of z is not 0 dB. In fact, the mean of x is

$$\mu_x = \mu_x^{(1)} = \exp\left(\mu_y + \frac{\sigma_y^2}{2}\right) \tag{A.10}$$

If that mean is to be unity, then

$$\mu_y = -\frac{\sigma_y^2}{2} \qquad \text{or, from (A.3),} \qquad \mu_z = -0.1 \cdot \ln(10) \cdot \frac{\sigma_z^2}{2} \tag{A.11}$$

Alternatively, if you do take the mean of y to be zero, then be aware that, from (A.10), the mean of x is greater than one and the average of the received power (A.9) is

$$P_{av} = P_0 \cdot \exp\left(\frac{\sigma_y^2}{2}\right) \tag{A.12}$$

APPENDIX B: A JAKES-LIKE METHOD
OF COMPLEX GAIN GENERATION

General Description

Simulations of fading channels need a method of generating channel complex gain samples that is statistically reliable, yet computationally undemanding. Jakes' method has earned considerable popularity because of its simplicity. In order to produce a Gaussian process with statistics corresponding to isotropic scattering, it mimics the model Jakes used to calculate the power spectrum in the first place - rays arriving uniformly from all directions with the same power.

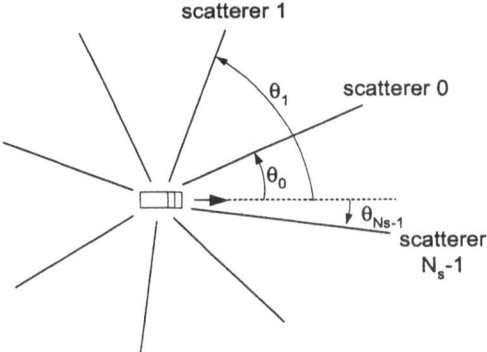

Computationally, the method is very attractive: the processing load per sample is reasonable; its structure lends itself to real time DSP implementation; it can be run forward or backward in time; and the time step size can be varied at will. Its statistical properties have had a more checkered history, particularly in regards to the creation of multiple uncorrelated generators. The original algorithm [**Jake74**] had a problem with persistent correlation between supposedly independent generators. The modification in [**Dent93**] got around part of the difficulty, but introduced other forms of correlation, as well as non-WSS behaviour, which together make it unsuitable for simulation.

In this appendix, we introduce a new Jakes-like generator that solves the problems of the previous versions and has good statistical behaviour: it is a good approximation to a Gaussian random process, it is bandlimited and wide-sense stationary, and it is easy to create multiple uncorrelated generators. We pay for this progress by requiring more rays than the earlier algorithms to achieve the desired power spectrum (or autocorrelation function).

We assume that there are N_s scatterers equispaced in azimuth around the mobile (their distance is irrelevant). All scattered signals have the same amplitude, though their phases ϕ_i are random. If the mobile is translated a distance x, and the angle of scatterer i is θ_i with respect to the direction of movement, then the change of phase in signal i can be written variously as

$$2 \cdot \pi \cdot \frac{x}{\lambda} \cdot \cos\left(\theta_i\right) = 2 \cdot \pi \cdot \frac{v \cdot t}{\lambda} \cdot \cos\left(\theta_i\right) = 2 \cdot \pi \cdot f_D \cdot t \cdot \cos\left(\theta_i\right) \tag{B1}$$

The generated complex gain is just the sum of the scattered signals

$$g(t) = \frac{1}{\sqrt{N_s}} \cdot \sum_{i=0}^{N_s-1} \exp\left[j \cdot \left(\phi_i + 2 \cdot \pi \cdot f_D \cdot t \cdot \cos\left(\theta_i\right) \right) \right]$$

$$= \frac{1}{\sqrt{N_s}} \cdot \sum_{i=0}^{N_s-1} \exp\left[j \cdot \left(\phi_i + \omega_i \cdot t \right) \right] \tag{B2}$$

where $\omega_i = 2\pi f_D \cos(\theta_i)$.

Statistical Characterization

Next, we check the statistical behaviour of the generator. We should use time-averaged or space-averaged statistics here, since the only point at which ensemble statistics enter is the initialization of ϕ_n, and ensemble statistics are not meaningful if we intend to collect simulation statistics from a long run using a single sample function.

Recall that the lowpass complex envelope $g(t)$ represents a real bandpass process (the radio frequency signal we receive when the transmitted signal is unmodulated carrier). This process will be wide-sense stationary if the real and imaginary components of $g(t)$ are uncorrelated and have the same autocorrelation functions. This gives us our first test, whether

$$\text{Avg}_t(g(t) \cdot g(t - \tau)) = 0 \tag{B3}$$

where Avg_t denotes a time average. This is not the same as the autocorrelation function, since there is no conjugate on one of the factors. Substitute (B2), and you obtain

$$\text{Avg}_t(g(t) \cdot g(t - \tau)) = \frac{1}{N_s} \cdot \sum_{i=0}^{N_s-1} \sum_{n=0}^{N_s-1} e^{j \cdot (\phi_i + \phi_n)} \cdot \text{Avg}_t\left[e^{j \cdot (\omega_i + \omega_n) \cdot t} \right] \cdot e^{-j \cdot \omega_n \cdot \tau} \tag{B4}$$

It is clear that the average will be zero, as desired, provided:

* none of the $\omega_n = 0$

* no ω_i is the negative of another ω_n

Since the Doppler shifts ω_i are determined by the arrival angles, we next consider conditions on N_s and θ_0 in

$$\theta_i = \theta_0 + \frac{2 \cdot \pi}{N_s} \cdot i \qquad 0 \le i \le N_s - 1 \tag{B5}$$

to ensure the required zero average. The conditions are, essentially,

* number of rays N_s is odd

* no arrival angle equals $\pi/2$ or $-\pi/2$

The procedure below does the checks for us (ε is a "closeness" threshold and the procedure returns the value 1 if it's too close to non-WSS behaviour):

$$
\text{nonWSS}\left(\theta_0, N_s, \varepsilon\right) :=
\begin{array}{|l}
\text{flag} \leftarrow 0 \\[4pt]
\text{flag} \leftarrow 1 \quad \text{if} \ \ \text{mod}\left(N_s, 2\right) = 0 \\[4pt]
\text{for} \ i \in 0 .. N_s - 1 \\[4pt]
\qquad \begin{array}{|l}
\theta \leftarrow \theta_0 + \dfrac{2 \cdot \pi}{N_s} \cdot i \\[10pt]
\text{flag} \leftarrow 1 \quad \text{if} \ \left| \theta - \dfrac{\pi}{2} \right| < \varepsilon \\[10pt]
\text{flag} \leftarrow 1 \quad \text{if} \ \left| \theta + \dfrac{\pi}{2} \right| < \varepsilon
\end{array} \\[30pt]
\text{flag}
\end{array}
\qquad \text{(B6)}
$$

As our second statistical test, we want the autocorrelation to resemble the Bessel function J_0 and have variance $1/2$ (so that it doesn't change the signal power passing through the channel). That is, we want

$$
R_g(\tau) = \text{Avg}_t\left(\overline{g(t) \cdot g(t-\tau)}\right) = J0\left(2 \cdot \pi \cdot f_D \cdot \tau\right) \qquad \text{(B7)}
$$

Substitute (B2) to check the validity of (B7):

$$
\text{Avg}_t\left(\overline{g(t) \cdot g(t-\tau)}\right) = \frac{1}{N_s} \cdot \sum_{i=0}^{N_s-1} \sum_{n=0}^{N_s-1} e^{j \cdot \left(\phi_i - \phi_n\right)} \cdot \text{Avg}_t\left[e^{j \cdot \omega_i \cdot t} \cdot e^{-j \cdot \omega_n \cdot (t-\tau)}\right]
$$

$$
= \frac{1}{N_s} \cdot \sum_{i=0}^{N_s-1} \sum_{n=0}^{N_s-1} e^{j \cdot \left(\phi_i - \phi_n\right)} \cdot \delta(i,n) \cdot e^{j \cdot \omega_n \cdot \tau}
$$

$$\blacksquare = \frac{1}{N_s} \cdot \sum_{n=0}^{N_s-1} e^{j \cdot \omega_n \cdot \tau} = \frac{1}{N_s} \cdot \sum_{n=0}^{N_s-1} e^{j \cdot 2 \cdot \pi \cdot f_D \cdot \tau \cdot \cos\left(\theta_n\right)}$$

$$\blacksquare \approx J0\left(2 \cdot \pi \cdot f_D \cdot \tau\right) \tag{B8}$$

since the summation in the second last line is a discrete approximation to an integral that is one definition of the J_0 function, given here as

$$J0\left(2 \cdot \pi \cdot f_D \cdot \tau\right) = \frac{1}{2 \cdot \pi} \cdot \int_{-\pi}^{\pi} \exp\left[j \cdot 2 \cdot \pi \cdot f_D \cdot \tau \cdot \cos(\theta) \right] d\theta \tag{B9}$$

The approximation is quite good for just a few arrivals. To check, use (B5), (B8) to define

$$R_g\left(u, N_s, \theta_o\right) := \frac{1}{N_s} \cdot \sum_{n=0}^{N_s-1} e^{j \cdot 2 \cdot \pi \cdot u \cdot \cos\left(\frac{2 \cdot \pi}{N_s} \cdot n + \theta_o\right)} \tag{B10}$$

where $u=f_D\tau$. Also define the error $R_e\left(u, N_s, \theta_o\right) := \left(R_g\left(u, N_s, \theta_o\right)\right) - J0(2 \cdot \pi \cdot u)$

Plot the error and the desired autocorrelation (the J0 function) for a reasonable number of arrivals and two different offset angles θ_0:

$$u := 0, 0.1 .. 15 \qquad N_s := 15 \quad \text{(keep it odd)}$$

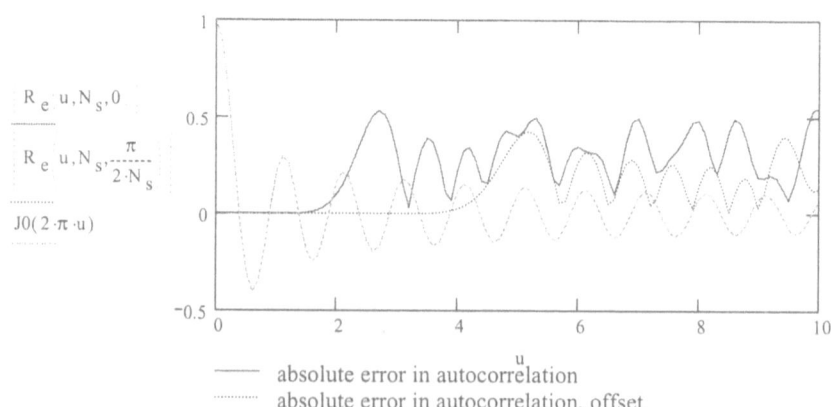

R_e u,N_s,0   —— absolute error in autocorrelation

R_e u,N_s,$\frac{\pi}{2 \cdot N_s}$ ··········· absolute error in autocorrelation, offset

J0(2·π·u)   ········· ideal autocorrelation

The quality of the approximation depends on the offset angle θ_0. You see the best and the worst cases above, with values 0 and $\pi/2N_s$. The approximation is quite good out to about $u=2$ to 4 for $N_s=15$ rays, and we can approximately double the interval of good fit by doubling the number of rays. Since very few receivers have a memory extending over more than 2 or 3 multiples of $1/f_D$, we can have confidence that the simulated results will reflect the desired Doppler spectrum.

It should be noted that all Jakes-like generators - ones with a finite number of discrete rays - show this autocorrelation behaviour for large values of $u=f_D\tau$, since they all make use of a discrete approximation to J_0. However, the original algorithm [**Jake74**] and the modified algorithm [**Dent93**] have an interval of good representation that approximately twice as large as the method proposed here, for the same number of rays. We have paid a price in computational load for good WSS behaviour.

The Jakes Generator Functions

We're now ready to create some procedures for generating complex gain sequences. We just mimic the equations above, using a subscript J to avoid naming conflicts when this appendix is used as an include file.

The first pair of procedures gives us a single gain process. The intialization prepares an array G with the state variables of the generator: the first half of G holds the randomized phases ϕ_n, and the second half holds the Doppler shifts, also randomized by selection of the first arrival angle θ_0. Both phases and angles include useful constants. If the check for WSS properties fails, it simply picks another θ_0.

$$N_s := 15 \qquad \text{(remember to make it odd)}$$

$$\text{Jakes\_init(dummy)} := \begin{array}{|l} \text{for } i \in 0 .. N_s - 1 \\[2mm] \quad G_i \leftarrow j \cdot \text{rnd}(2\cdot\pi) \\[2mm] \quad \theta_0 \leftarrow \text{rnd}\!\left(\dfrac{2\cdot\pi}{N_s}\right) \\[2mm] \text{while } \text{nonWSS}\!\left(\theta_0, N_s, 0.01\right) \\[2mm] \quad \theta_0 \leftarrow \text{rnd}\!\left(\dfrac{2\cdot\pi}{N_s}\right) \\[2mm] \text{for } i \in N_s .. 2\cdot N_s - 1 \\[2mm] \quad G_i \leftarrow -j \cdot 2\cdot\pi\cdot\cos\!\left(\theta_0 + \dfrac{2\cdot\pi}{N_s}\cdot i\right) \\[2mm] G \end{array} \qquad\qquad \text{(B11)}$$

The second function generates the gain samples in a computationally efficient form equivalent to (B2) by

$$\text{Jakes\_gen}(u, G) := \frac{1}{\sqrt{N_s}} \cdot \sum_{i=0}^{N_s - 1} \exp\left(G_i + u \cdot G_{i+N_s}\right) \tag{B12}$$

where $u = x/\lambda = f_D t$. As an example of use,

$$G := \text{Jakes\_init}(1) \qquad \text{Jakes\_gen}(0.3, G) = -0.028 + 0.906j$$

Here's a sample function. To see different sample functions, place the cursor on the G initialization and recompute by pressing F9.

$$iJ := 0 .. 100 \qquad g_{iJ} := \text{Jakes\_gen}(0.1 \cdot iJ, G)$$

Multiple Complex Gain Generators

Having demonstrated that individual sample functions have the desired statistical behaviour, we'll look at the issue of creating several uncorrelated complex gain sequences. We'd want them if we were simulating channels with delay spread (for the separate TDL taps) or with diversity reception.

The method is very simple: give the different generators different values of first arrival angle θ_0. Since the generators then have no arrival angles (hence Doppler shifts) in common, their cross correlation will be zero, as required. Let's check it. First, define the processes:

$$g1(t) = \frac{1}{\sqrt{N_s}} \cdot \sum_{i=0}^{N_s - 1} \exp\left[j \cdot \left(\phi1_i + \omega1_i \cdot t\right)\right] \qquad g2(t) = \frac{1}{\sqrt{N_s}} \cdot \sum_{i=0}^{N_s - 1} \exp\left[j \cdot \left(\phi2_i + \omega2_i \cdot t\right)\right]$$

$$\tag{B13}$$

Their cross correlation must be zero. It is calculated like autocorrelation:

$$\text{Avg}_t\left(g1(t)\cdot\overline{g2(t-\tau)}\right) = \frac{1}{N_s}\cdot\sum_{i=0}^{N_s-1}\sum_{k=0}^{N_s-1}e^{j\cdot(\phi1_i-\phi2_k)}\cdot\text{Avg}_t\left[e^{j\cdot\omega_i\cdot t}\cdot e^{-j\cdot\omega_k\cdot(t-\tau)}\right]$$

$$=\frac{1}{N_s}\cdot\sum_{i=0}^{N_s-1}\sum_{k=0}^{N_s-1}e^{j\cdot(\phi1_i-\phi2_k)}\cdot e^{j\cdot\omega_k}\cdot\text{Avg}_t\left[e^{-j\cdot(\omega_k-\omega_i)\cdot t}\right]$$

$$(B14)$$

If the Doppler shifts ω_i and ω_k are all different, then the average is zero, as desired. Whether they are different depends on the offset arrival angles. These angles are in the range $0\le\theta1_0,\theta2_0<\dfrac{2\cdot\pi}{N_s}$ by design, and in order to have the Doppler shifts all different, we simply need to ensure

$$\theta1_0\ne\theta2_0 \qquad\qquad (B15)$$

Multiple Jakes Generator Functions

Next, we'll use the Jakes generator functions above to create several uncorrelated complex gain sequences. All you do is use several instances with different randomly selected timing phases and arrival angles:

$$G1 := \text{Jakes\_init}(1) \qquad\qquad G2 := \text{Jakes\_init}(1)$$

$$(B16)$$

$$g1_{iJ} := \text{Jakes\_gen}(0.1\cdot iJ, G1) \qquad g2_{iJ} := \text{Jakes\_gen}(0.1\cdot iJ, G2)$$

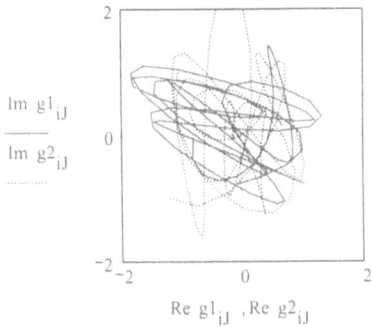

For rough work - visualization or simple demos - this is sufficient. However, for proper statistical work, you should also check that the generators are not correlated due to an unfortunate choice of arrival angles. Here's a test for arrival angles too close together:

$$\text{corrgen}\left(G_1, G_2, \varepsilon\right) := \left(\left|G_{1N_s} - G_{2N_s}\right| < \varepsilon\right)$$

Try it for the generators G1 and G2 above,

$$\text{corrgen}(G1, G2, 0.01) = 0 \qquad \text{You'll see 0 if they are all right, 1 if they're too correlated}$$

We can create N_g generators randomly, relying on the checks above. The procedure below forms a G matrix, with each column representing a different generator. If a generator is too closely correlated with those already selected, the procedure tries another random first arrival angle.

$$\text{multi\_Jakes\_init}\left(N_g, \text{dummy}\right) := \begin{vmatrix} G^{<0>} \leftarrow \text{Jakes\_init(dummy)} \\ \text{for } k \in 1..N_g - 1 \\ \quad \begin{vmatrix} G^{<k>} \leftarrow \text{Jakes\_init(k)} \\ \text{while } \sum_{i=0}^{k-1} \text{corrgen}\left(G^{<i>}, G^{<k>}, 0.01\right) > 0 \\ \quad G^{<k>} \leftarrow \text{Jakes\_init(k)} \end{vmatrix} \\ G \end{vmatrix}$$

And here's how to use it... We'll make two uncorrelated processes:

$$N_g := 2 \qquad\qquad G := \text{multi\_Jakes\_init}\left(N_g, 1\right)$$

$$g1_{iJ} := \text{Jakes\_gen}\left(0.1 \cdot iJ, G^{<0>}\right) \qquad g2_{iJ} := \text{Jakes\_gen}\left(0.1 \cdot iJ, G^{<1>}\right)$$

and plot the two processes on the same graph below.

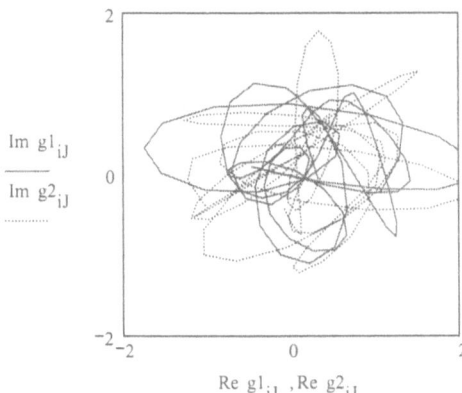

Put the cursor on the
initialization and press F9
to see other pairs of sample
functions

APPENDIX C: VISUALIZATION OF A RANDOM STANDING WAVE

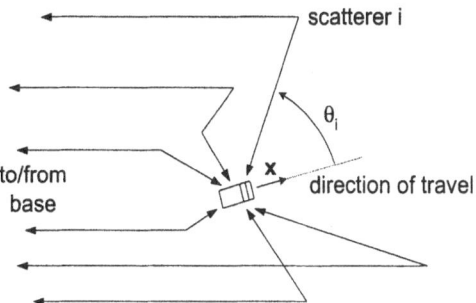

This example illustrates a typical random spatial distribution of signal power. Assume that we have N := 15 rays crossing a region of space in which position is given in rectangular coordinates by the complex variable x. (See sketch). The angles of arrival with respect to the direction of travel and the phases at x=0 (a reference position) are given by

$$i := 0 .. N - 1$$

$$\theta_i := rnd(2 \cdot \pi) \qquad \text{random arrival angles}$$

$$\phi_i := rnd(2 \cdot \pi) \qquad \text{random phases at } x{=}0$$

At some point x, the change of path length along ray i, compared to x=0, is given by

$$-Re\left(x \cdot e^{-j \cdot \theta_i}\right)$$

For convenience, use unit wavelength, so that $\lambda := 1$ and $\beta := \dfrac{2 \cdot \pi}{\lambda}$. Therefore the field as a function of position x is given by the superposition of the ray signals

$$E(x) := \frac{1}{\sqrt{N}} \cdot \sum_i e^{j \cdot \phi_i} \cdot \exp\left(-j \cdot \beta \cdot Re\left(x \cdot e^{-j \cdot \theta_i}\right)\right)$$

To see what it looks like, we'll set up a rectangular grid 2λ x 2λ

$$M := 20 \qquad m := 0 .. M \qquad n := 0 .. M \qquad x_{m,n} := 0.1 \cdot \lambda \cdot (m + j \cdot n)$$

then calculate the composite signal values and their squared magnitude for the plot

$$y_{m,n} := E\left(x_{m,n}\right) \qquad\qquad ymag_{m,n} := \left(\left| y_{m,n} \right|\right)^2$$

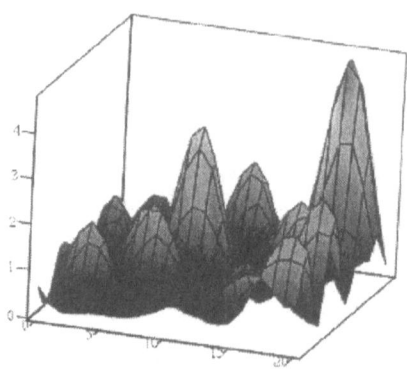

Spatial Variation of Signal Power

ymag

Drag this surface around to different orientations with your mouse.

This is only a 2λ x 2λ patch, so you can imagine the rapid fluctuations in signal strength experienced by the mobile as it drives through.

You can see quasiperiodicity (i.e., it's similar to, but not quite, periodic behaviour).

If you want to see other sample functions, go back up to the equations for ϕ and θ and recalculate them (put the cursor on them and press F9) and view the plots again. Alternatively, just go the menu bar and select Math/Calculate Worksheet.

APPENDIX D: ANIMATION OF THE COMPLEX GAIN

By now, you have probably seen the random standing wave example in Appendix C, and you know that when the mobile moves through the standing wave, the result is rapid changes in amplitude and phase of the received signal. You can gain considerable insight into the effect on signal detection of these changes just by watching them - and that's what this worksheet does for you. Of course, it has a serious purpose, but it's also interesting. So get out the popcorn, sit back, and enjoy...

Complex Gain Theatre

We're going to see the complex gain as a function of space (time), leaving a trail behind it. We'll need the Jakes complex gain generator, so include it.

⊡ Reference:D:\COURSES\MobChann\paperbook\Jakesgen.mcd(R)

unit wavelength:	steps per wavelength	step size

$\lambda := 1$ $N_\lambda := 20$ $\Delta\lambda := \dfrac{\lambda}{N_\lambda}$

make the trail one wavelength long:

and the frame to frame step $\lambda/4$ long:

$i := 0 .. N_\lambda$ $step := floor\left(\dfrac{N_\lambda}{4}\right)$

$G := Jakes\_init(1)$ $g_i := Jakes\_gen((FRAME \cdot step + i) \cdot \Delta\lambda , G)$

Here's how you do it (you could also read the Mathcad manual here). From the menu, View Animate. In the dialog box, select the FRAME range from 0 to 40 at 5 frames/sec. Next use your mouse to enclose the graph below in a dotted selection box, then click the Animate button in the dialog box. After an interval of computation, during which you see a thumbnail of the animation, Mathcad pops up a window with your animation. You can resize the window if you like, then play from beginning to end, or step it with the slider.

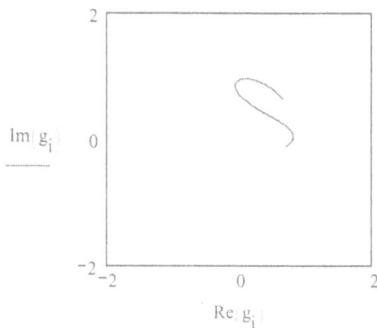

To see other examples of complex gain trajectories, put your cursor on the initialization equation above, press F9, then reaimate.

Click here for a precalculated animation.

You should be noting a few points:

* The quasi-periodicity shows up as a looping, or Lissajous-like, behaviour.

* Most of the time, the amplitude is reasonably good, but sometimes the gain speeds right past the origin.

* When the gain goes past the origin, the phase changes very, very quickly. In fact, the closer it gets to the origin, the faster the phase changes. The receiver then has double trouble, with a phase hit just at the point of poorest SNR. This means big problems for phase locked loops and for detection methods that rely on phase or its derivative, such as a discriminator circuit.

These points are strikingly evident if we plot dB magnitude and the phase derivative against position in space.

$$dB(x) := if\left(x < 10^{-14}, -140, 10 \cdot \log(x)\right)$$

$$i := 0 .. 40 \cdot step$$

$$g\,dB_i := dB\left[\left(|\,Jakes\_gen(i \cdot \Delta\lambda, G)\,|\right)^2\right]$$

$$\Delta\phi_i := arg\left(\overline{Jakes\_gen(i \cdot \Delta\lambda, G) \cdot Jakes\_gen((i-1) \cdot \Delta\lambda, G)}\right) \qquad x_i := i \cdot \Delta\lambda$$

Animate the two graphs below by enclosing them *both* in the selection box. Use the same FRAME parameters as earlier. If you want to see different version, just put your cursor on the Jakes initialization earlier, then come back here and reanimate.

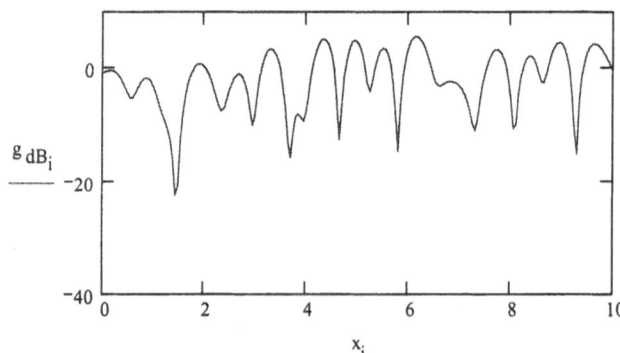

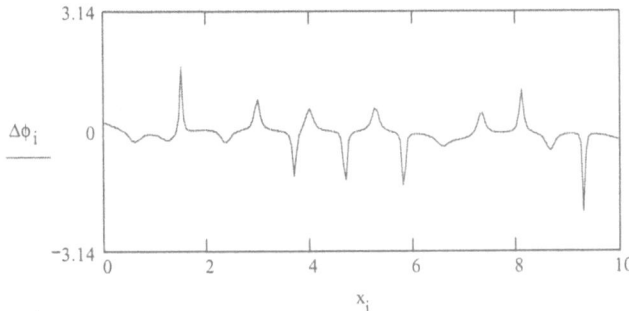

Click here for a
precalculated
animation.

It is clear from the animations that peaks of phase derivative coincide with deep fades, just as we deduced from the polar plots.

APPENDIX E: ANIMATION OF TIME VARYING FREQUENCY AND IMPULSE RESPONSES

More theatrics and special effects. This one brings frequency selective fading to life, as nulls sweep through the frequency band and group delay twitches back and forth - all because the mobile is moving through the random field pattern. It certainly makes one appreciate the challenges facing designers (possibly you?) of receive modems that operate over wide bandwidths.

⊡ Reference:D:\COURSES\MobChann/ebook\Units.mcd(R)

⊡ Reference:D:\COURSES\MobChann/ebook\Jakesgen.mcd(R)

Choose the bandwidth, Doppler spread and sampling rate:

$$W := \frac{300 \cdot kHz}{2} \qquad f_D := 100 \cdot Hz \qquad f_s := 4 \cdot W \qquad t_s := f_s^{-1}$$

Assume a linear decay in the delay power profile:

$$\tau_d := 10 \cdot \mu s \qquad P_g(\tau) := \frac{2}{\tau_d} \cdot \left(1 - \frac{\tau}{\tau_d}\right) \cdot (0 \cdot \mu s \le \tau) \cdot (\tau < \tau_d)$$

$$\tau := 0 \cdot \mu s, 0.1 \cdot \mu s .. 2 \cdot \tau_d$$

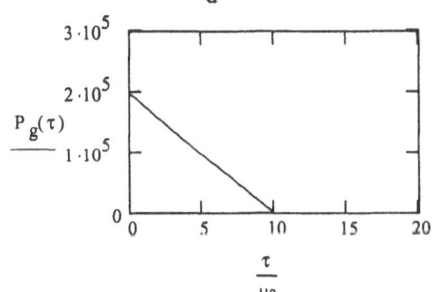

so the area $\sigma_g^2 = 1$

number of taps in the model:

$$N_t := \text{ceil}\left(\frac{\tau_d}{t_s}\right) \qquad n := 0 .. N_t - 1$$

$$N_t = 6$$

standard dev'n per tap:

$$\sigma_{g_n} := \sqrt{\int_{n \cdot t_s}^{(n+1) \cdot t_s} P_g(\tau) \, d\tau}$$

Assume isotropic scattering, so use Jakes' model for generation:

Initialize the parameter blocks of the coefficients:

$$g^{<n>} := \text{Jakes\_init}(n)$$

At any observation time it_s, coefficient n is given by:

$$\sigma_{g_n} \cdot Jakes\_gen\left(f_D \cdot t_s \cdot i, g^{<n>}\right)$$

and the frequency response at observation time it_s is

$$G(i,f) := \sum_n \sigma_{g_n} \cdot Jakes\_gen\left(f_D \cdot t_s \cdot i, g^{<n>}\right) \cdot e^{-j \cdot 2 \cdot \pi \cdot f \cdot t_s \cdot n}$$

Group delay of an LTI filter is normally defined as

$$\tau_g = -\frac{1}{2 \cdot \pi} \cdot \frac{d}{d f} arg(G(f))$$

We'll use the logarithmic derivative to calculate the group delay as function of frequency, and we'll make it time varying:

$$D(i,f) := -\frac{1}{2 \cdot \pi} \cdot Im\left[\frac{\sum_n -j \cdot 2 \cdot \pi \cdot t_s \cdot n \cdot \sigma_{g_n} \cdot Jakes\_gen\left(f_D \cdot t_s \cdot i, g^{<n>}\right) \cdot e^{-j \cdot 2 \cdot \pi \cdot f \cdot t_s \cdot n}}{G(i,f)}\right]$$

We'll plot the magnitude and group delay with a frequency resolution of $1/8\tau_d$ and observation times separated by

$$N_{obs} := 200 \qquad samples$$

$$N_f := ceil\left(2 \cdot W \cdot 8 \cdot \tau_d\right) \qquad N_f = 24 \qquad \text{number of frequency samples}$$

$$k := 0 .. N_f$$

$$f_k := -W + k \cdot \frac{1}{8 \cdot \tau_d}$$

$$H_k := G\left(FRAME \cdot N_{obs}, f_k\right) \qquad H := 3 \cdot \left|\overrightarrow{H}\right| \qquad \text{scale the magnitude so that it and delay spread can share the vertical axis}$$

$$del_k := D\left(FRAME \cdot N_{obs}, f_k\right)$$

Here's how you make the graph move (you could also read the Mathcad manual here). From the menu, View Animate. In the dialog box, select the FRAME range from 0 to 20 at 5 frames/sec. Next use your mouse to enclose the graph below in a dotted selection box, then click the Animate button in the dialog box. After an interval of computation, during which you see a thumbnail of the animation, Mathcad pops up a window with your animation. You can resize the window if you like, then play from beginning to end, or step it with the slider.

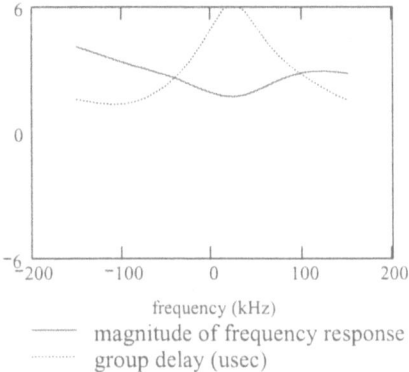

frequency (kHz)

——— magnitude of frequency response

·········· group delay (usec)

Try other sample functions by putting your cursor on the Jakes initialization above and pressing F9 (or from menu Math/Calculate Worksheet), then reanimating the graph.

Click here for a precalculated animation.

A surprising feature of this animation is the sudden reversals of group delay in certain bands.

Now for the impulse response. Like the frequency response, it's complex, but here we'll just look at its magnitude.

$$h_n := \sigma \cdot g_n \cdot Jakes\_gen\left(f_D \cdot t_s \cdot N_{obs} \cdot FRAME, g^{<n>}\right)$$

$$h := \overrightarrow{h}$$

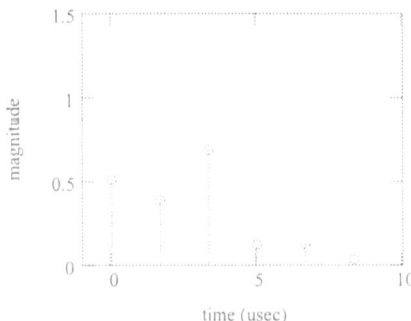

magnitude

time (usec)

Animate this one the same way as the frequency response. It's a little less informative, but it does give you an idea of how quickly the impulse response changes, and therefore how fast the receive modem must adapt.

Click here for a precalculated animation.

APPENDIX F: JOINT SECOND ORDER STATISTICS

Most of the material in this section will already be familiar to you from your first course on random processes, back when you were an undergraduate. It's a little more elaborate here, but only because the processes are complex. Reference material is in [**Papo84, Proa95**], among many others. This is not an *exciting* section, but it is useful.

The discussion below focuses on the flat fading gain process $g(t)$ and its associated power spectrum $S_g(\nu)$ and autocorrelation function $R_g(\Delta t)$. However, it can be adapted wholesale to the instantaneous frequency response process $H(f)$ and its associated autocorrelation function $C_g(\Delta f)$ and delay power profile $P_g(\tau)$. This could be important to you at some point, so remember it!

Symmetries

We have a random process $g(t)$. Assuming that it's WSS, its autocorrelation function is

$$R_g(\Delta t) = \frac{1}{2} \cdot E\left(\overline{g(t) \cdot g(t - \Delta t)}\right) \tag{F1}$$

and we have the transform pair

$$S_g(\nu) = \int_{-\infty}^{\infty} R_g(\Delta t) \cdot e^{-j \cdot 2 \cdot \pi \cdot \nu \cdot \Delta t} \, d\Delta t \qquad R_g(\Delta t) = \int_{-\infty}^{\infty} S_g(\nu) \cdot e^{j \cdot 2 \cdot \pi \cdot \nu \cdot \Delta t} \, d\nu \tag{F2}$$

Because the power spectrum $S_g(\nu)$ is real, the autocorrelation function $R_g(\Delta t)$ is conjugate symmetric

$$R_g(-\Delta t) = \overline{R_g(\Delta t)} \tag{F3}$$

that is, its real part is even and its imaginary part is odd. If the power spectrum is even (i.e., symmetric about $\nu=0$), as well as real, then $R_g(\Delta t)$ is real, not just conjugate symmetric. We saw an example of the in **Section 5.1** with the idealized symmetric Doppler spectrum and the real Bessel function autocorrelation.

In general, though, the power spectrum is not even. Consider, for example, a carrier frequency offset between transmitter and receiver, one that has not been corrected by AFC. In this case, the channel output is

$$y(t) = e^{j \cdot 2 \cdot \pi \cdot f_0 \cdot t} \cdot g(t) \cdot s(t) \tag{F4}$$

where f_0 is the frequency offset. Here we might wish to modify the definition of the complex gain process to include the exponential, so that we could continue to write $y(t) = g(t)s(t)$. If so, the power spectrum might look like the left hand figure below:

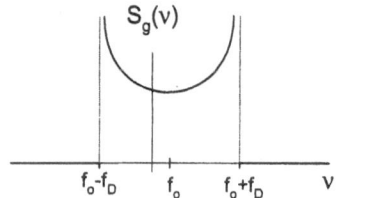
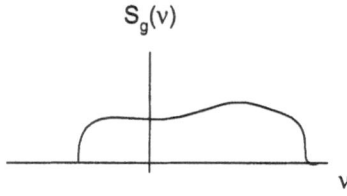

and we would have

$$R_g(\Delta t) = \sigma_g^2 \cdot e^{j \cdot 2 \cdot \pi \cdot f_0 \cdot \Delta t} \cdot J_0 (2 \cdot \pi \cdot f_D \cdot \Delta t) \tag{F5}$$

which is obviously complex, but it remains conjugate symmetric. More generally still, the Doppler spectrum could be aymmetric, as in the right hand figure above, if the scatterers are not isotropic. In this case the autocorrelation function is conjugate symmetric, but not of the tidy form of (F5); we would simply have an even real part and an odd imaginary part:

$$\text{Re}(R_g(-\Delta t)) = \text{Re}(R_g(\Delta t)) \qquad \text{Im}(R_g(-\Delta t)) = -\text{Im}(R_g(\Delta t)) \tag{F6}$$

Moments and Series Expansions

From **(5.1.5) and (5.1.9)**, we have the familiar relation

$$\int_{-\infty}^{\infty} S_g(v) \, dv = R_g(0) = \sigma_g^2 \tag{F7}$$

Since the power spectrum is real and non-negative, if we normalize it by its area (F7), it becomes similar to a pdf, in that we can define moments. The first moment is the mean frequency offset

$$v_m = \frac{1}{\sigma_g^2} \cdot \int_{-\infty}^{\infty} v \cdot S_g(v) \, dv = \frac{1}{j \cdot 2 \cdot \pi \cdot \sigma_g^2} \cdot R'_g(0) \tag{F8}$$

where the second equality follows from differentiation of $R_g(\Delta t)$ in (F2). We can define a second moment similarly

$$v_m^{(2)} = \frac{1}{\sigma_g^2} \cdot \int_{-\infty}^{\infty} v^2 \cdot S_g(v) \, dv = \frac{-1}{4 \cdot \pi^2 \cdot \sigma_g^2} \cdot R''_g(0) \tag{F9}$$

but it is more useful to work with the second *central* moment (like a variance). The rms bandwidth is defined by its square (the mean square bandwidth)

$$v_{rms}^2 = \frac{1}{\sigma_g^2} \cdot \int_{-\infty}^{\infty} (v - v_m)^2 \cdot S_g(v)\, dv = v_m^{(2)} - v_m^2$$

$$= \frac{1}{4 \cdot \pi^2 \cdot \sigma_g^2} \cdot \left[-R''_g(0) + \frac{1}{\sigma_g^2} \cdot R'_g(0)^2 \right] \qquad \text{(F10)}$$

In the case that the mean frequency offset is zero, so that $R'_g(0)=0$, we have a simple and very useful relation

$$v_{rms}^2 = \frac{-1}{4 \cdot \pi^2 \cdot \sigma_g^2} \cdot R''_g(0) \qquad \text{(F11)}$$

Why is it useful? Because we now have a simple series expansion for the autocorrelation

$$R_g(\Delta t) = R_g(0) + \frac{1}{2} \cdot R''(0) \cdot \Delta t^2 + \dots.$$

$$= \sigma_g^2 \cdot \left(1 - 2 \cdot \pi^2 \cdot v_{rms}^2 \cdot \Delta t^2 + \dots \right) \qquad \text{(F12)}$$

which is sufficient to describe the decorrelation due to fading in many detection schemes, such as differential detection. This decorrelation produces an error floor, or irreducible bit error rate, even if there is no channel noise. That is, no increase in transmitted power can bring the error rate below the floor. Because of (F12), the error floor usually varies quadratically with rms Doppler spread.

For static frequency selective transmission, the same argument can be applied in the other domain pair to produce the series for the spaced frequency correlation function

$$C_g(\Delta f) = \sigma_g^2 \cdot \left(1 - 2 \cdot \pi^2 \cdot \tau_{rms}^2 \cdot \Delta f^2 + \dots \right) \qquad \text{(F13)}$$

after the time axis has been shifted so that the mean delay $\tau_m = 0$.

Correlations With the Derivative

The derivative $g'(t)$ is important in characterizing random FM effects, fade rates and durations, etc. It is also Gaussian, since differentiation is a linear operation. We need the auto- and cross-correlations to characterize the process fully. The sketch below show it in linear systems terms:

From elementary random process theory (filtering of random processes), the second order characterizations of $g'(t)$ are

$$S_{g'}(v) = (|j \cdot 2 \cdot \pi \cdot v|)^2 \cdot S_g(v) = (2 \cdot \pi \cdot v)^2 \cdot S_g(v) \tag{F14}$$

$$R_{g'}(\Delta t) = -\frac{d^2}{d \Delta t^2} R_g(\Delta t) = \frac{1}{2} \cdot E\left(g'(t) \cdot \overline{g'(t - \Delta t)}\right) \tag{F15}$$

from which (or by use of (F11)) we also obtain for a process with zero mean frequency offset

$$R_{g'}(0) = \sigma_{g'}^2 = 4 \cdot \pi^2 \cdot \sigma_g^2 \cdot v_{rms}^2 \tag{F16}$$

The power in the derivative is proportional to the rms bandwidth - interesting! The cross power density and cross correlation functions are

$$S_{gg'}(v) = -j \cdot 2 \cdot \pi \cdot v \cdot S_g(v) \tag{F17}$$

$$R_{gg'}(\Delta t) = -\frac{d}{d \Delta t} R_g(\Delta t) = \frac{1}{2} \cdot E\left(g(t) \cdot \overline{g'(t - \Delta t)}\right) \tag{F18}$$

and the cross correlation, evaluated at $\Delta t = 0$, is proportional to the mean frequency offset

$$R_{gg'}(0) = \sigma_{gg'}^2 = -j \cdot 2 \cdot \pi \cdot \sigma_g^2 \cdot v_m \tag{F19}$$

From (F19), we see that a process and its derivative are uncorrelated when evaluated at the same time, provided the mean frequency offset is zero, as in a symmetric power spectrum.

Probability Density Functions

Armed with all these auto and cross correlation functions, we find it easy to write the joint probability density functions. As a reminder (we saw this in Section 4.2 as **(4.2.2)**)

$$P_g(g) = \frac{1}{2 \cdot \pi \cdot \sigma_g^2} \cdot \exp\left[-\frac{1}{2} \cdot \frac{(|g|)^2}{\sigma_g^2}\right] \tag{F20}$$

Similarly, for the derivative

$$P_{g'}(g') = \frac{1}{2 \cdot \pi \cdot \sigma_{g'}^2} \cdot \exp\left[-\frac{1}{2} \cdot \frac{(|g'|)^2}{\sigma_{g'}^2}\right] \tag{F21}$$

For the joint pdf of separated samples of $g(t)$, we have to work a little harder. Define the vector

$$\mathbf{g} = \begin{bmatrix} g(t) \\ g(t-\tau) \end{bmatrix} = \begin{bmatrix} g_1 \\ g_2 \end{bmatrix} \qquad g_1 \text{ and } g_2 \text{ are abbreviations} \tag{F22}$$

and covariance matrix, with related quantities, as

$$S = \frac{1}{2} \cdot E\left(\mathbf{g} \cdot \mathbf{g}^T\right) = \frac{1}{2} \cdot E\left[\begin{bmatrix} \left(|g_1|\right)^2 & g_1 \overline{g_2} \\ \overline{g_1} g_2 & \left(|g_2|\right)^2 \end{bmatrix}\right] = \begin{bmatrix} R_g(0) & R_g(\tau) \\ R_g(\tau) & R_g(0) \end{bmatrix} \tag{F23}$$

$$\det(S) = R_g(0)^2 - \left(|R_g(\tau)|\right)^2 \qquad S^{-1} = \frac{1}{\det(S)} \cdot \begin{bmatrix} R_g(0) & -R_g(\tau) \\ -R_g(\tau) & R_g(0) \end{bmatrix} \tag{F24}$$

The pdf for complex Gaussian vectors is given by

$$P_\mathbf{g}(\mathbf{g}) = \frac{1}{4 \cdot \pi^2 \cdot \det(S)} \cdot \exp\left(-\frac{1}{2} \cdot \overline{\mathbf{g}}^T \cdot S^{-1} \cdot \mathbf{g}\right) \tag{F25}$$

If you don't have this form in your introductory text on communications, see [**Proa95**] on multivariate Gaussian distributions for real vectors. For complex vectors, we recognize that each element has two components, real and imaginary, that are independent and identically distributed. Substitution into the real multivariate distribution doubles the dimensionality, but use of complex notation reduces it to (F25). In scalar form, our bivariate distribution of complex gains becomes

$$P_{g1g2}(g_1, g_2) = \frac{1}{4 \cdot \pi^2 \cdot \left[R_g(0)^2 - \left(|R_g(\tau)|\right)^2\right]} \times \blacksquare$$

$$\blacksquare \times \exp\left[-\frac{R_g(0) \cdot \left[\left(|g_1|\right)^2 + \left(|g_2|\right)^2\right] - 2 \cdot \mathrm{Re}\left(R_g(\tau) \cdot \overline{g_1} g_2\right)}{2 \cdot \left[R_g(0)^2 - \left(|R_g(\tau)|\right)^2\right]}\right] \tag{F26}$$

Similarly, for the joint pdf of $g(t)$ and its derivative $g'(t)$, we adopt the alternative definitions of vector

$$\mathbf{g} = \begin{bmatrix} g(t) \\ g'(t-\tau) \end{bmatrix} \tag{F27}$$

and covariance matrix, with related quantities, as

$$
S = \frac{1}{2} \cdot E\left(g \cdot g^{\overline{T}}\right) = \frac{1}{2} \cdot E\left[\begin{bmatrix} (|g|)^2 & \overline{g \cdot g'} \\ \overline{g \cdot g'} & (|g'|)^2 \end{bmatrix}\right] = \begin{bmatrix} R_g(0) & R_{gg'}(\tau) \\ R_{gg'}(\tau) & R_{g'}(0) \end{bmatrix}
$$

(F28)

$$
\det(S) = R_g(0) \cdot R_{g'}(0) - \left(\left| R_{gg'}(\tau) \right|\right)^2 \qquad S^{-1} = \frac{1}{\det(S)} \cdot \begin{bmatrix} R_{g'}(0) & -R_{gg'}(\tau) \\ -R_{gg'}(\tau) & R_g(0) \end{bmatrix}
$$

(F29)

$$
P_{gg'}(g, g') = \frac{1}{4 \cdot \pi^2 \cdot \left[R_g(0) \cdot R_{g'}(0) - \left(\left| R_{gg'}(\tau) \right|\right)^2 \right]} \times \blacksquare
$$

$$
\blacksquare \times \exp\left[-\frac{R_{g'}(0) \cdot (|g|)^2 - 2 \cdot \mathrm{Re}\left(R_{gg'}(\tau) \cdot \overline{g \cdot g'}\right) + R_g(0) \cdot (|g'|)^2}{2 \cdot \left[R_g(0) \cdot R_{g'}(0) - \left(\left| R_{gg'}(\tau) \right|\right)^2 \right]} \right]
$$

(F30)

If the spectrum is symmetric, so that the mean frequency offset is zero, then $g(t)$ and $g'(t)$ are uncorrelated and S is diagonal. That makes the joint pdf just the product of the two individual pdfs given in (F20) and (F21). With (F7) and (F16),

$$
P_{gg'}(g, g') = \frac{1}{4 \cdot \pi^2 \cdot \sigma_g^2 \cdot \sigma_{g'}^2} \cdot \exp\left[-\frac{1}{2} \cdot \left[\frac{(|g|)^2}{\sigma_g^2} + \frac{(|g'|)^2}{\sigma_{g'}^2} \right] \right]
$$

(F31)

APPENDIX G: JOINT PROBABILITY DENSITY FUNCTIONS
IN POLAR COORDINATES

In **Appendix F**, we obtained the joint pdf of samples of the complex gain and its derivative in rectangular coordinates. Here we obtain their counterparts in polar coordinates. We use

$$g = g_r + j \cdot g_i = r \cdot e^{j \cdot \psi} \qquad \text{with Jacobian r:} \quad r \cdot dr \cdot d\psi = dg_r \cdot dg_i \qquad (G1)$$

Pdf of Gain

First, the pdf of g itself, in Cartesian and polar coordinates (the familiar **Rayleigh pdf**):

$$P_g(g) = \frac{1}{2 \cdot \pi \cdot \sigma_g^2} \cdot \exp\left[-\frac{1}{2} \cdot \frac{(|g|)^2}{\sigma_g^2} \right] \qquad P_{r\psi}(r, \psi) = \frac{r}{2 \cdot \pi \cdot \sigma_g^2} \cdot \exp\left(-\frac{1}{2} \cdot \frac{r^2}{\sigma_g^2} \right) \qquad (G2)$$

Joint Pdf of Spaced Samples of Gain

Next, the joint pdf of spaced samples of g. We use

$$g_1 = g(t) = r_1 \cdot e^{j \cdot \psi_1} \qquad g_2 = g(t - \tau) = r_2 \cdot e^{j \cdot \psi_2} \qquad \text{Jacobian} \qquad r_1 \cdot r_2 \qquad (G3)$$

From **Appendix F**, we have the Cartesian form (F26), which we convert to polars, using the two Jacobians:

$$P_{r1\psi1r2\psi2}(r_1, \psi_1, r_2, \psi_2) =$$

$$= \frac{r_1 \cdot r_2}{4 \cdot \pi^2 \cdot \sigma_g^4 \left[1 - (|\rho|)^2 \right]} \cdot \exp\left[-\frac{r_1^2 + r_2^2 - 2 \cdot r_1 \cdot r_2 \cdot \mathrm{Re}\left[\rho \cdot e^{j \cdot (\psi_2 - \psi_1)} \right]}{2 \cdot \sigma_g^2 \left[1 - (|\rho|)^2 \right]} \right]$$

$$(G4)$$

where $R_g(0) = \sigma_g^2$ and the complex correlation coefficient ρ is defined by $R_g(\tau) = \rho \cdot R_g(0)$.

Joint Pdf of Spaced Samples of Amplitude

We obtain the joint pdf of r_1 and r_2 by integrating (G4) over ψ_1 and ψ_2:

$$P_{r1r2}(r_1,r_2)=\int_0^{2\cdot\pi}\int_0^{2\cdot\pi}P_{r1\psi1r2\psi2}(r_1,\psi_1,r_2,\psi_2)\,d\psi_1\,d\psi_2$$

$$\blacksquare=\frac{r_1\cdot r_2}{\sigma_g^4\cdot\left[1-(|\rho|)^2\right]}\cdot\exp\left[-\frac{r_1^2+r_2^2}{2\cdot\sigma_g^2\cdot\left[1-(|\rho|)^2\right]}\right]\times\blacksquare$$

$$\blacksquare\times\int_0^{2\cdot\pi}\int_0^{2\cdot\pi}\frac{1}{4\cdot\pi^2}\cdot\exp\left[\frac{2\cdot r_1\cdot r_2\cdot|\rho|\cdot\cos\left(\alpha+\psi_2-\psi_1\right)}{2\cdot\sigma_g^2\cdot\left[1-(|\rho|)^2\right]}\right]d\psi_1\,d\psi_2$$

$$(G5)$$

where $\alpha=arg(\rho)$. Integration over ψ_1 gives the Bessel function and integration over ψ_2 gives 2π. Therefore

$$P_{r1r2}(r_1,r_2)=\frac{r_1\cdot r_2}{\sigma_g^4\cdot\left[1-(|\rho|)^2\right]}\cdot\exp\left[-\frac{r_1^2+r_2^2}{2\cdot\sigma_g^2\cdot\left[1-(|\rho|)^2\right]}\right]\cdot I_0\left[\frac{2\cdot r_1\cdot r_2\cdot|\rho|}{2\cdot\sigma_g^2\cdot\left[1-(|\rho|)^2\right]}\right]$$

$$(G6)$$

Joint Pdf of Spaced Samples of Phase

The joint pdf of ψ_1 and ψ_2 is obtained by integrating (G4) over r_1 and r_2:

$$P_{\psi1\psi2}(\psi_1,\psi_2)=\int_0^\infty\int_0^\infty P_{r1\psi1r2\psi2}(r_1,\psi_1,r_2,\psi_2)\,dr_1\,dr_2 \qquad (G7)$$

The double integral is evaluated in [**Dave87**]. If we define

$$\beta=|\rho|\cdot\cos\left(\alpha+\psi_2-\psi_1\right)$$

$$(G8)$$

where $\alpha = arg(\rho)$, then

$$P_{\psi 1 \psi 2}(\Psi_1, \Psi_2) = \frac{1 - (|\rho|)^2}{4 \cdot \pi^2} \cdot \frac{(1 - \beta)^2 + \beta \cdot (\pi - acos(\beta))}{(1 - \beta)^{1.5}} \tag{G9}$$

Pdf of Derivative

We can represent the derivative $g'(t)$ in polar coordinates as

$$g' = s \cdot e^{j \cdot \theta} \tag{G10}$$

Since it, too, is Gaussian we immediately have

$$P_{s\theta}(s, \theta) = \frac{s}{2 \cdot \pi \cdot \sigma_{g'}^2} \cdot exp\left(-\frac{1}{2} \cdot \frac{s^2}{\sigma_{g'}^2}\right) \quad \text{where, from (F16), } \sigma_{g'}^2 = 4 \cdot \pi^2 \cdot \sigma_g^2 \cdot v_{rms}^2 \tag{G11}$$

Joint Pdf of Complex Gain and Derivative

We can also represent the derivative $g'(t)$ in terms of the derivatives of the polar coordinate representation of $g(t)$ (note that they are evaluated at the same time t):

$$g = g_r + j \cdot g_i = r \cdot cos(\psi) + j \cdot r \cdot sin(\psi)$$

$$g' = g'_r + j \cdot g'_i$$

$$= (r' \cdot cos(\psi) - r \cdot \psi' \cdot sin(\psi)) + j \cdot (r' \cdot sin(\psi) + r \cdot \psi' \cdot cos(\psi)) \tag{G12}$$

To convert the pdf in Cartesian coordinates g_r, g_i, g'_r, g'_i to the coordinates r, ψ, r', ψ', we need the Jacobian. Take differentials in (G12) and calculate the determinant of the 4x4 matrix (symbolically, if you are using Mathcad to do it) and the resulting Jacobian is just r^2. The pdf in Cartesians is given as (F31) in **Appendix F** for symmetric spectra. Substitution of (G12) and use of the Jacobian gives

$$P_{r\psi r'\psi'}(r, \psi, r', \psi') = \frac{r^2}{4 \cdot \pi^2 \cdot \sigma_g^2 \cdot \sigma_{g'}^2} \cdot exp\left[-\frac{1}{2} \cdot \left(\frac{r^2}{\sigma_g^2} + \frac{r^2 \cdot \psi'^2 + r'^2}{\sigma_{g'}^2}\right)\right] \tag{G13}$$

Remember that this is the joint density of $g(t)$ and $g'(t)$ for symmetric spectra (less general than $g(t)$ and $g'(t-\tau)$).

We can obtain pdfs of various subsets of variables from (G13) by integration (the symbolic processor in Mathcad is a time-saver), and we can obtain conditional densities, such as the pdf o r' and y', given r, by dividing (G13) by the appropriate unconditional pdfs. We'll do a couple of them below.

Joint Pdf of Amplitude and Its Derivative

By integration of (G13), we have

$$P_{rr'}(r,r') = \int_{-\infty}^{\infty} \int_{0}^{2\cdot\pi} P_{r\psi r'\psi'}(r,\psi,r',\psi') \, d\psi \, d\psi'$$

$$= \int_{-\infty}^{\infty} \frac{r^2}{2\cdot\pi\cdot\sigma_g^2 \cdot\sigma_{g'}^2} \cdot \exp\left[-\frac{1}{2}\cdot\left(\frac{r^2}{\sigma_g^2} + \frac{r^2\cdot\psi'^2 + r'^2}{\sigma_{g'}^2}\right)\right] d\psi' \qquad (G14)$$

Symbolic integration gives

$$P_{rr'}(r,r') = \frac{r}{\sqrt{2\cdot\pi}\cdot\sigma_g^2 \cdot\sigma_{g'}} \cdot \exp\left[-\frac{1}{2}\cdot\left(\frac{r^2}{\sigma_g^2} + \frac{r'^2}{\sigma_{g'}^2}\right)\right] \qquad (G15)$$

This pdf is useful in calculation of level crossing rates.

Joint Pdf of Derivatives of Polar Representation of Gain

We will integrate (G13) with respect to r and y to obtain the joint pdf of r' and y'.

$$P_{r'\psi'}(r',\psi') = \int_{0}^{\infty} \int_{0}^{2\cdot\pi} \frac{r^2}{4\cdot\pi^2\cdot\sigma_g^2 \cdot\sigma_{g'}^2} \cdot \exp\left[-\frac{1}{2}\left(\frac{r^2}{\sigma_g^2} + \frac{r^2\cdot\psi'^2 + r'^2}{\sigma_{g'}^2}\right)\right] d\psi \, dr$$

$$= \int_{0}^{\infty} \frac{r^2}{2\cdot\pi\cdot\sigma_g^2 \cdot\sigma_{g'}^2} \cdot \exp\left[-\frac{1}{2}\cdot\left(\frac{r^2}{\sigma_g^2} + \frac{r^2\cdot\psi'^2 + r'^2}{\sigma_{g'}^2}\right)\right] dr \qquad (G16)$$

Symbolic integration and simplification gives

$$P_{r'\psi'}(r',\psi') = \frac{\sigma_{g'} \cdot \sigma_g}{2 \cdot \sqrt{2 \cdot \pi}} \cdot \frac{\exp\left(-\frac{1}{2} \cdot \frac{r'^2}{\sigma_{g'}^2}\right)}{\left(\sigma_{g'}^2 + \sigma_g^2 \cdot \psi'^2\right)^{1.5}} \tag{G17}$$

Since (G17) can be written in product form, the derivatives r' and y' are independent.

Pdf of Phase Derivative

The pdf of y' is obtained by integrating out r' in (G17)

$$P_{\psi'}(\psi') = \int_{-\infty}^{\infty} P_{r'\psi'}(r',\psi')\, dr' = \frac{1}{2} \cdot \frac{\sigma_g \cdot \sigma_{g'}^2}{\left(\sigma_{g'}^2 + \sigma_g^2 \cdot \psi'^2\right)^{1.5}} \tag{G18}$$

This expression is useful in analysis of random FM.

APPENDIX H: GENERATION OF COMPLEX GAIN FOR IMPORTANCE SAMPLING

The functions in this Appendix can save you a *lot* of simulation time, whether you use them directly in Mathcad simulations, or transcribe them to C or Matlab. They let you generate a vector of equispaced complex gain samples that conforms exactly to the autocorrelation function you supply. No approximation by discrete rays, as in the Jakes method. No approximate spectra, as in the conventional filtered noise method. It is exact. Also - and this is really the point - you can specify the variance of the centre sample. If you make it equal to the nominal variance of the process, then the gain vector is just like that produced by the usual methods (apart from the fact that the autocorrelation is exact). However, if you make it smaller than the nominal variance, you generate fades in the middle of the vectors, so you can use importance sampling methods.

The idea is simple: pick a value for the complex gain sample in the middle of the vector. Then "grow" the rest of the gain vector by adding a sample to the left of the midpoint, then a sample to the right of the midpoint. Now we have three samples. Add a sample to the left of this trio, then to the right, giving five. And so on. How do we ensure the right statistics? The mean value of each sample is obtained by linear estimation from the previously generated samples, and we add to it a sample of random noise having the right estimation error variance.

First, we get the prediction coefficients and error variances by Levinson's recursion [**Hayk96**].

$$
\text{levinson}(\mathbf{r}) := \begin{vmatrix}
N \leftarrow \text{length}(\mathbf{r}) \\[4pt]
a_{0,0} \leftarrow 1 \\[4pt]
E_0 \leftarrow r_0 \\[4pt]
\text{for } n \in 1 .. N-1 \\[4pt]
\quad \begin{vmatrix}
\mathbf{a}^{<n>} \leftarrow \mathbf{a}^{<n-1>} \\[4pt]
a_{n,n} \leftarrow 0 \\[4pt]
D_{n-1} \leftarrow \displaystyle\sum_{i=0}^{n} \overline{r_i} \cdot a_{n-i,n} \\[4pt]
\gamma_n \leftarrow \dfrac{D_{n-1}}{E_{n-1}} \\[4pt]
E_n \leftarrow \left[1 - \left(\left| \gamma_n \right| \right)^2 \right] \cdot E_{n-1} \\[4pt]
\text{for } i \in 0 .. n \\[4pt]
\quad t_i \leftarrow a_{i,n} - \gamma_n \cdot \overline{a_{n-i,n}} \\[4pt]
\mathbf{a}^{<n>} \leftarrow \mathbf{t}
\end{vmatrix} \\[4pt]
\mathbf{a} \leftarrow \text{augment}(\text{augment}(\mathbf{a}, E), \gamma)
\end{vmatrix}
$$

This procedure is based on methods described in **[Proa95, Orfa85]** for real processes and in **[Hayk96]** for complex processes. The argument r is a vector of samples of the autocorrelation function at increasing lags. For example,

$$f_D := 100 \qquad \Delta t := 200 \cdot 10^{-6} \qquad \text{(Doppler in Hz and spacing in seconds)}$$

$$N := 5 \qquad i := 0 .. N - 1 \qquad r_i := J0\left(2 \cdot \pi \cdot f_D \cdot \Delta t \cdot i\right)$$

Note that the length N of the vector must be odd. Levinson's recursion gives

$$A := \text{levinson}(r) \qquad A = \begin{bmatrix} 1 & 1 & 1 & 1 & 1 & 1 & 0 \\ 0 & -0.996 & -1.99 & -2.986 & -3.982 & 7.872 \cdot 10^{-3} & 0.996 \\ 0 & 0 & 0.998 & 2.984 & 5.963 & 3.106 \cdot 10^{-5} & -0.998 \\ 0 & 0 & 0 & -0.998 & -3.978 & 1.225 \cdot 10^{-7} & 0.998 \\ 0 & 0 & 0 & 0 & 0.998 & 4.828 \cdot 10^{-10} & -0.998 \end{bmatrix}$$

The first N columns of A are the prediction error filters of increasing order (column i corresponds to order i). Therefore the prediction coefficients of order i are the negatives of the elements 1 to i of column i. Column N contains the corresponding prediction error variances of increasing order, and column $N+1$ contains the PARCOR coefficients.

Next, we need a method of generating complex Gaussian noise of unit variance. Easy:

$$\text{cgauss}(x) := \sqrt{-2 \cdot \ln(\text{rnd}(1))} \cdot \exp(j \cdot \text{rnd}(2 \cdot \pi))$$

You will probably recognize it from the discussion on the Rayleigh pdf. First, $-2 \cdot \ln(\text{rnd}(1))$ generates an exponentially distributed variate with mean 2. Recall that the squared magnitude of a unit variance complex Gaussian variate has this distribution. The square root converts it to Rayleigh magnitude, and the complex exponential gives it uniformly distributed phase around the circle.

The gain generation procedure is on the next page. The arguments are A, an array created by the Levinson procedure, and σ_c, the standard deviation of the centre sample. The loop repeatedly creates first the left sample, then the right sample. The computation goes up linearly with each time through the loop, since the length of the inner product increases. Here you may want to make a small change. If accuracy of the autocorrelation is of less importance than computation, you could alter the A matrix before using the gain generator as follows: the high order prediction columns peter out as you go down the column, so set the very small values to zero. Mathcad will skip any complex multiplication in the summation sign when it discovers a zero value for the $A_{k,i}$ or $A_{k,i+1}$ multiplicand.

$$\text{IS\_gen}(\mathbf{A}, \sigma_c) := \begin{vmatrix} N \leftarrow \text{length}(\mathbf{A}^{<0>}) \\ \text{mid} \leftarrow \dfrac{N-1}{2} \\ g_{\text{mid}} \leftarrow \sigma_c \cdot \text{cgauss}(0) \\ \text{for } i \in 1..\text{mid} \\ \quad \begin{vmatrix} g_{\text{mid}-i} \leftarrow -\displaystyle\sum_{k=1}^{2 \cdot i - 1} A_{k, 2 \cdot i - 1} \cdot g_{\text{mid}-i+k} + \sqrt{A_{i,N}} \cdot \text{cgauss}(i) \\ g_{\text{mid}+i} \leftarrow -\displaystyle\sum_{k=1}^{2 \cdot i} A_{k, 2 \cdot i} \cdot g_{\text{mid}+i-k} + \sqrt{A_{i+1,N}} \cdot \text{cgauss}(i) \end{vmatrix} \\ g \end{vmatrix}$$

The examples below show that each time you use IS_gen, it produces a new gain vector, and that puts a fade in the middle. In constrast, the unbiased generation (σ_c=0.5) does not constrain the trajectory. **To see the effect, recalculate by putting the cursor on the equation and pressing F9 repeatedly.**

$$g_{\text{unbias}} := \text{IS\_gen}(\mathbf{A}, 0.5) \qquad\qquad g_{\text{bias}} := \text{IS\_gen}(\mathbf{A}, 0.1)$$

$$g_{\text{unbias}} = \begin{bmatrix} 1.401 - 0.141i \\ 1.18 - 0.218i \\ 0.961 - 0.301i \\ 0.742 - 0.39i \\ 0.521 - 0.483i \end{bmatrix} \qquad\qquad g_{\text{bias}} = \begin{bmatrix} -0.511 - 0.09i \\ -0.256 - 0.044i \\ 0.011 + 0.01i \\ 0.281 + 0.069i \\ 0.542 + 0.129i \end{bmatrix}$$

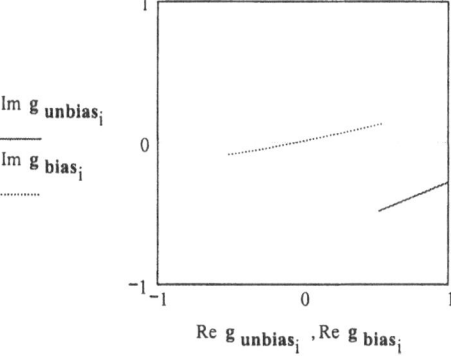

Im g_{unbias_i}
—————
Im g_{bias_i}
....

Re g_{unbias_i} , Re g_{bias_i}

APPENDIX I: LINEAR TIME VARIANT FILTERS

We have seen that the fundamental model of fading is a linear time-varying (LTV) filter. You may not be as familiar with the input-output relations for LTV filters as you are for their time invariant (LTI) counterparts. Since LTV filters play such a large role in mobile communications, this section is provided to give you some familiarity with them. We'll start by putting the LTI filter in context.

Linear Time Invariant Systems

A fundamental description of a linear system is the convolution relation between input and output:

$$y(t) = \int_{-\infty}^{\infty} h(\tau) \cdot s(t - \tau) \, d\tau \tag{I.1}$$

Note that we are using densities, for continuously distributed scatterers, with integrals. Our earlier expressions in the physical models of **Section 3.1** focused on discrete scatterers, with summations; however, they can be regenerated by appropriate use of impulses in the integrals. We interpret the impulse response $h(\tau)$ as the gain, at any observation time t, experienced by an input signal delayed by τ. We have two times of interest here: the observation time t and the input time t-τ; equivalently, the two times of interest are the observation time and the delay. In LTI systems, the impulse response depends only on the delay.

Complex exponentials - the Fourier basis - play a key role here, since they are eigenfunctions of convolution, in the sense that they are self-reproducing. If we transform the impulse response in the delay domain τ (sometimes called the echo delay domain) to obtain $H(f)$, with inverse transform

$$h(\tau) = \int_{-\infty}^{\infty} H(f) \cdot e^{j \cdot 2 \cdot \pi \cdot f \cdot \tau} \, df \tag{I.2}$$

then (3.3.1) takes the form

$$y(t) = \int_{-\infty}^{\infty} s(t - \tau) \cdot \int_{-\infty}^{\infty} H(f) \cdot e^{j \cdot 2 \cdot \pi \cdot f \cdot \tau} \, df \, d\tau \tag{I.3}$$

$$= \int_{-\infty}^{\infty} H(f) \cdot S(f) \cdot e^{j \cdot 2 \cdot \pi \cdot f \cdot t} \, df \tag{I.4}$$

In particular, if $s(t) = exp(j2\pi f_0 t)$ is a complex exponential at frequency f_0, then $S(f) = \delta(f - f_0)$ is an impulse in the frequency domain and

$$y(t) = H(f_0) \cdot e^{j \cdot 2 \cdot \pi \cdot f_0 \cdot t} \tag{I.5}$$

so that $H(f_o)$ is the gain experienced at any observation time t by a carrier offset by frequency f_o.

The impulse response and frequency response are both fundamental system descriptions, and are equivalent, in the sense that either can be obtained from the other through Fourier transformation. Note that the Fourier exponentials have infinite duration and constant frequency - this Fourier frequency is not the same as the instantaneous frequency of a time varying signal, as measured perhaps by a discriminator circuit.

Linear Time Varying Systems

In an LTV system, the basic descriptions are more elaborate, because more complicated phenomena are present. For example, a modulator produces frequencies not present in the input signal, something not possible in an LTI system. You have already seen the delay-Doppler spread function in **Section 3.2**, but the most common descriptions are the time variant impulse response and the time variant transfer function. For more detailed discussions, see [**Bell63, Bell64, Kenn69, Stee92, Proa95**], of which [**Stee92**] is the best recent one.

The *time variant impulse response* $g(t,\tau)$ is the gain at observation time t experienced by a signal delayed by τ (alternatively, the response at time t to an impulse at t-τ). You saw this one as **(3.3.6)** in Section 3.3. The equivalent of convolution is

$$y(t) = \int_{-\infty}^{\infty} g(t,\tau) \cdot s(t-\tau) \, d\tau \tag{I.6}$$

Note that the impulse response is now a function of two variables: the observation time and the delay. We can Fourier transform it in either or both domains to obtain the frequency makeup of either or both of the output or the input, respectively. The former is related to the Doppler spread, and the latter to the delay spread.

The *time variant transfer function* results if we transform the impulse response in the delay domain, with inverse transform:

$$g(t,\tau) = \int_{-\infty}^{\infty} G(t,f) \cdot e^{j \cdot 2 \cdot \pi \cdot f \cdot \tau} \, d\alpha \tag{I.7}$$

Substitution into (I.6) gives

$$y(t) = \int_{-\infty}^{\infty} s(t-\tau) \cdot \int_{-\infty}^{\infty} G(t,f) \cdot e^{j \cdot 2 \cdot \pi \cdot f \cdot \tau} \, df \, d\tau$$

$$= \int_{-\infty}^{\infty} S(f) \cdot G(t,f) \cdot e^{j \cdot 2 \cdot \pi \cdot f \cdot t} \, df \tag{I.8}$$

In particular, if the input is a complex exponential in time at frequency f_o (an impulse in frequency) we have

$$y(t) = G\left(t, f_0\right) \cdot e^{j \cdot 2 \cdot \pi \cdot f_0 \cdot t} \tag{I.9}$$

That is, the time-variant transfer function $G(t, f_0)$ is the gain at observation time t experienced by a carrier that is offset by frequency f_0. It may seem strange to mix time and frequency like this, but remember that f is related to the delay domain, not the observation time.

The two less common characterizations are obtained by transformation in the observation time domain, that is, projecting $y(t)$ onto its own Fourier basis with frequency β. If we transform (I.8), we have

$$Y(\beta) = \int_{-\infty}^{\infty} e^{-j \cdot 2 \cdot \pi \cdot \beta \cdot t} \cdot \int_{-\infty}^{\infty} S(f) \cdot G(t, f) \cdot e^{j \cdot 2 \cdot \pi \cdot f \cdot t} \, df \, dt$$

$$= \int_{-\infty}^{\infty} S(f) \cdot \int_{-\infty}^{\infty} G(t, f) \cdot e^{j \cdot 2 \cdot \pi \cdot f \cdot t} \cdot e^{-j \cdot 2 \cdot \pi \cdot \beta \cdot t} \, dt \, df$$

$$= \int_{-\infty}^{\infty} S(f) \cdot \Gamma(\beta - f, f) \, df \tag{I.10}$$

where $\Gamma(v, f)$ is termed the *output Doppler-spread function*. Here's why: if the input is a complex exponential at frequency f_0, then $S(f) = \delta(f - f_0)$ and

$$Y(\beta) = \Gamma\left(\beta - f_0, f_0\right) \quad \text{or} \quad Y\left(f_0 + v\right) = \Gamma\left(v, f_0\right) \tag{I.11}$$

where $v = \beta - f_0$ is the difference, or Doppler, frequency. This simple relation says that the component of the output at Fourier frequency $f_0 + v$ is $\Gamma(v, f_0)$ times the component of the input at Fourier frequency f_0. It exposes the ability of time varying systems to produce frequencies not present in the input. This function is not as esoteric as it seems. The multitone modulation format OFDM (orthogonal frequency division multiplexing) [**Salt67, Cimi85**] maintains orthogonality between the tones, or subchannels, on a static channel. On a fast fading channel, the distortion of the signal causes interchannel interference (ICI): a response in one subchannel to modulation in a nearby subchannel.

The last of the four characterizations at least has the advantage of familiarity, from **Section 3.2** Like the output Doppler-spread function, it uses transformation in the observation time domain. We define

$$g(t, \tau) = \int_{-\infty}^{\infty} \gamma(v, \tau) \cdot e^{j \cdot 2 \cdot \pi \cdot v \cdot t} \, dv \tag{I.12}$$

where $\gamma(v, \tau)$ is the *delay-Doppler spread function*. Substitution into (I.6) gives

$$y(t) = \int_{-\infty}^{\infty} \int_{-\infty}^{\infty} \gamma(v,\tau) \cdot s(t-\tau) \cdot e^{j \cdot 2 \cdot \pi \cdot v \cdot t} d\tau \, dv \qquad (I.13)$$

which we saw as **(3.2.10) and (3.2.11)** in Section 3.2. It shows that the delay-Doppler spread function is the gain experienced by $s(t-\tau)exp(j2\pi vt)$, a signal delayed by τ and Doppler shifted by v. In our mobile communications context, this one is interesting, because both domains v and τ are bounded by Doppler spread and delay spread respectively: we have $|v| \leq f_D$ and $0 \leq \tau \leq \tau_d$, so that (I.13) can be written as

$$y(t) = \int_{0}^{\tau_d} \int_{-f_D}^{f_D} \gamma(v,\tau) \cdot s(t-\tau) \cdot e^{j \cdot 2 \cdot \pi \cdot v \cdot t} d\tau \, dv \qquad (I.14)$$

With finite domains, there are several interesting possibilities for representation of $g(n,t)$, such as Fourier series, orthogonal polynomials or (if second order statistics are available) the Karhunen-Loeve expansion.

These four characterizations of time variant channels are equivalent, in the sense that they can be obtained from each other by Fourier transformation. Which one is most convenient depends on the problem at hand.

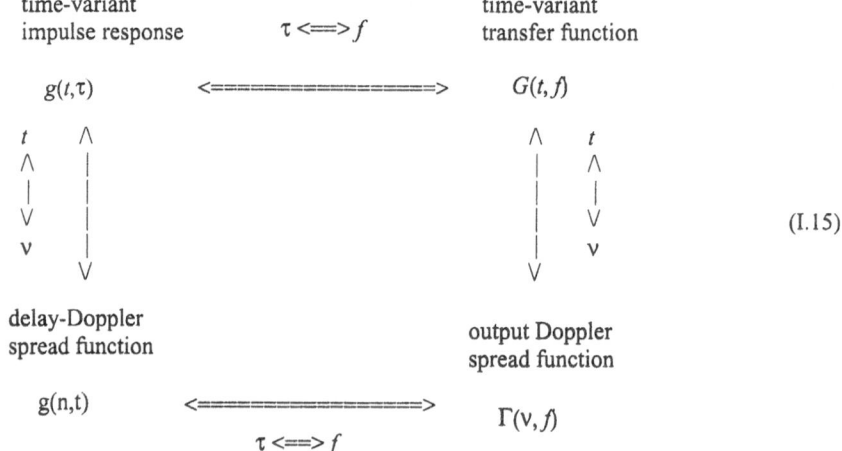

$$(I.15)$$

APPENDIX J: IS THE CHANNEL REALLY GAUSSIAN?
AN EXPERIMENT

We use the Gaussian (i.e., Rayleigh or Rice fading) model quite freely in design and analysis of mobile systems. It's a good testbed for modulation and detection - but does it reflect reality very well? This appendix lets you run some experiments to test the Gaussian approximation.

The variation in complex gain $g(t)$ is due primarily to the changing phases, rather than amplitudes, of the arrivals. Our experiment therefore consists of a large number of trials, each with randomly selected phases, but constant amplitudes. All we have to do then is make a histogram of the real and imaginary components and compare the result to the ideal pdf of the Gaussian model.

This is very close to real life, assuming that the selection of amplitudes is realistic. Of course, if the amplitudes are drawn from a Rayleigh pdf, the individual arrivals will all be Gaussian, and so will the sum. We will use an alternative model: the amplitudes have a uniform distribution about some mean value.

We'll start with your choice of number of arriving paths:

$$N_p := 5 \qquad\qquad i := 0 .. N_p - 1 \tag{J.1}$$

Next, give the arrivals some fixed amplitudes. We'll choose them from a uniform distribution with a mean of 1 and a width of $w := 1.2$ selectected in the range [0,2]

$$r_i := rnd(w) + 1 - \frac{w}{2} \qquad \text{<=== choose other values by putting the cursor on this} \tag{J.2}$$
$$\text{equation and pressing F9.}$$

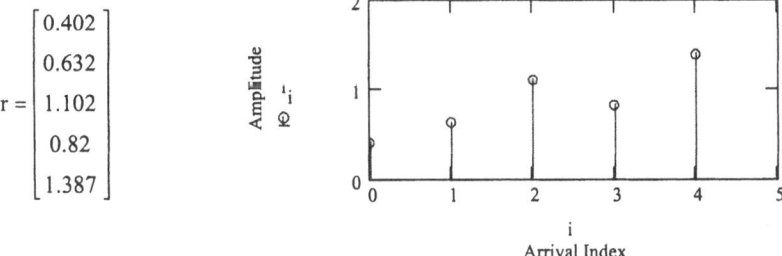

$$r = \begin{bmatrix} 0.402 \\ 0.632 \\ 1.102 \\ 0.82 \\ 1.387 \end{bmatrix}$$

We have to pick an interval size for the histogram. To be meaningful, it should be scaled to the standard deviation of the complex gain, a value that we can determine analytically. Each gain sample looks like this:

$$g = \sum_{i=0}^{N_p - 1} r_i \cdot e^{j \cdot \phi_i} \tag{J.3}$$

where the phases ϕ_i are independent among the paths and uniformly distributed over $[0,2\pi)$. Consequently, the standard deviations of the real part and the imaginary part are

$$\sigma_g := \sqrt{\frac{1}{2} \cdot \sum_{i=0}^{N_p - 1} \left(r_i\right)^2} \qquad \sigma_g = 1.479 \tag{J.4}$$

We'll take the histogram over the interval $[-h_{lim}\sigma_g, h_{lim}\sigma_g]$, that is, +/- h_{lim} standard deviations. You choose this parameter and the number of intervals in the histogram:

$$h_{lim} := 3 \qquad N_{int} := 29 \qquad k_{int} := 0 .. N_{int} - 1 \tag{J.5}$$

The interval width is then $\quad \Delta := \dfrac{2 \cdot h_{lim} \cdot \sigma_g}{N_{int}} \quad$ and the thresholds are given by

$$k_{th} := 0 .. N_{int} \qquad thresh_{k_{th}} := -h_{lim} \cdot \sigma_g + \Delta \cdot k_{th} \tag{J.6}$$

As a reference, we compute the histogram of an ideal Gaussian pdf of standard deviation σ_g:

$$hGauss_{k_{int}} := cnorm\left(\frac{thresh_{k_{int} + 1}}{\sigma_g}\right) - cnorm\left(\frac{thresh_{k_{int}}}{\sigma_g}\right) \tag{J.7}$$

and check that it sums to unity: $\Sigma hGauss = 0.997 \qquad$ It does (given the h_{lim} truncation).

Now to generate the gain samples. It's easiest to do this in a function, so you can choose the number of trials

$$gains\left(N_{trials}\right) := \begin{vmatrix} \text{for } n \in 0 .. N_{trials} - 1 \\ \quad \begin{vmatrix} \text{for } i \in 0 .. N_p - 1 \\ \quad \phi_i \leftarrow rnd(2 \cdot \pi) \\ \quad g_n \leftarrow \sum_{i=0}^{N_p - 1} r_i \cdot e^{j \cdot \phi_i} \end{vmatrix} \\ g \end{vmatrix} \tag{J.8}$$

All right, the preliminaries are over. We can get down to the experiment itself. Choose the number of trials and generate the gains:

$$N_{trials} := 10000 \qquad g := gains\left(N_{trials}\right) \tag{J.9}$$

Now get the histogram, normalized to sum to unity. Note that real and imaginary components are both histogrammed, so there are really $2N_{trials}$ elements.

$$h := \frac{\text{hist}(\text{thresh}, \text{Re}(g)) + \text{hist}(\text{thresh}, \text{Im}(g))}{2 \cdot N_{trials}} \tag{J.10}$$

The exact Gaussian histogram is shown faintly in cyan for reference.

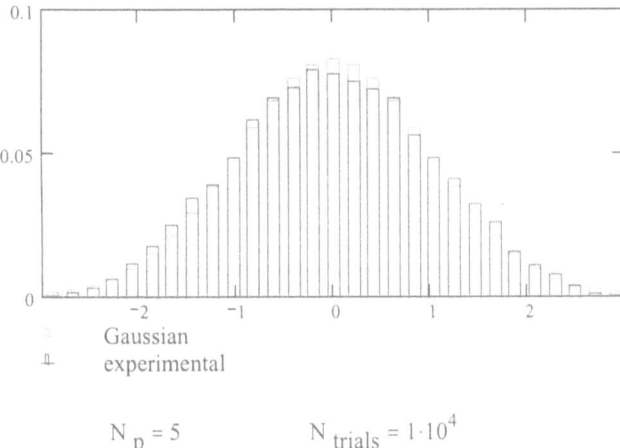

Gaussian
experimental

$$N_p = 5 \qquad\qquad N_{trials} = 1 \cdot 10^4$$

To do another run, put the cursor on the g= gains(N_{trials}) statement in (J.9) and press F9. To try a different set of amplitudes, go back to (J.2) and recalculate with F9. And of course, to try a different number of arrivals, change N_p in (J.1).

We expect to see discrepancies between the experiment and the Gaussian approximation out on the tails, since the true maximum amplitude is limited to the sum of the r_i. This is confirmed by the graphed results, particularly for a small number of paths, such as 3 or 5. Interestingly, though, the low amplitudes near the peak are also lower than Gaussian, even for larger numbers of paths, such as 9 or 10. This is true, to greater or lesser degree, regardless of the randomized set of amplitudes.

Our next check is the magnitude of the gain. We'll calculate a histogram of the magnitudes of many gain samples and compare it to the Rayleigh distribution. Here's the cumulative distribution function of a Rayleigh pdf with $\sigma^2 = 1$.

$$F_{Ray}(r) := 1 - \exp\left(\frac{-r^2}{2}\right) \tag{J.11}$$

We'll keep the same interval size as above, but use half as many intervals (since it's one-sided):

$$N_{int} := 15 \qquad\qquad k_{int} := 0 .. N_{int} - 1$$

$$k_{th} := 0 .. N_{int} \qquad\qquad \text{threshR}_{k_{th}} := \Delta \cdot k_{th}$$

and the histogram of an ideal Rayleigh variate, for reference, is

$$hRay_{k_{int}} := F_{Ray}\left(\frac{threshR_{k_{int}+1}}{\sigma_g}\right) - F_{Ray}\left(\frac{threshR_{k_{int}}}{\sigma_g}\right) \tag{J.12}$$

Now for the experiment. Choose the number of trials and generate the gains:

$$N_{trials} := 10000 \qquad g := gains\left(N_{trials}\right) \tag{J.13}$$

Now get the histogram of magnitudes, normalized to sum to unity.

$$h := \frac{hist\left(threshR, \overrightarrow{|g|}\right)}{N_{trials}} \tag{J.10}$$

The exact Rayleigh histogram is shown faintly in cyan for reference.

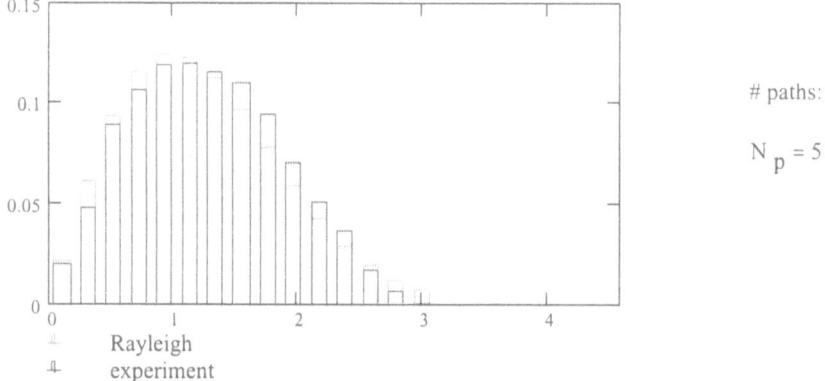

paths:

$$N_p = 5$$

To do another run, put the cursor on the g= gains(N$_{trials}$) statement in (J.13) and press F9. To try a different set of amplitudes, go back to (J.2) and recalculate with F9. And of course, to try a different number of arrivals, change N$_p$ in (J.1).

I'm sure you have noticed that, for relatively small numbers of paths, N_p=5, the Rayleigh model overestimates the small values. This is consistent with the Gaussian histogram above, since low values occur when both real and imaginary parts are small in magnitude.

What does this mean to radio link performance? Most communication problems occur in fades, and the probability of fades of various depths is a primary determinant of link performance. Since fades occur when the magnitude is small (i.e., both real and imaginary parts are small in magnitude), it's a little disquieting to see the Gaussian/Rayleigh model depart from reality in the small magnitude region. However, the model overestimates the probability of these events, so it is slightly pessimistic. Performance on real links should be a little better than the Gaussian model predicts. As for the tails of the pdf, the discrepancy has little practical significance, since they are not associated with fades or impaired performance.

APPENDIX K: FFT AND OVERLAP-ADD METHOD
OF COMPLEX GAIN GENERATION

General Comments

This appendix provides the procedures you need for generating complex gain sequences by filtering white noise, as described in **Section 9.2**. It uses an FFT for computational efficiency, so the procedure returns an array containing a segment of the complex gain process. However, its use of the overlap-add method of filtering [**Oppe89**] ensures that successive segments are continuous, so they impose no limit on the length of your simulation.

Our strategy in obtaining the gain generation procedure has several steps:

* Obtain the impulse response of the ideal filter for creating the process;

* Select a sampling rate and an appropriate block length, then use a Kaiser window to smooth the truncation of the impulse response;

* Use it in the overlap-add filtering of segments of white noise.

In contrast to the **Jakes method** of complex gain generation, filtering white noise produces a gain sequence that is both Gaussian and stationary, and it's easy to create multiple independent generators. On the other hand, it requires much more computation, and the power spectrum has some residual sidelobes due to truncation of the ideal impulse response. Your choice.

The Ideal Colouring Filter

Recall from **Section 5.1** that the Doppler power spectrum we are trying to mimic is

$$S_g(v) = \frac{\sigma_g^2}{\pi \cdot f_D} \cdot \frac{1}{\sqrt{1 - \left(\frac{v}{f_D}\right)^2}} \qquad \text{for} \quad |v| < f_D$$

It can be produced by feeding white noise with a flat power spectrum equal to σ_g^2 through a colouring filter with frequency response

$$H_n(v) = \frac{1}{\sqrt{\pi \cdot f_D}} \cdot \frac{1}{\left[1 - \left(\frac{v}{f_D}\right)^2\right]^{\frac{1}{4}}} \qquad \text{for} \quad |v| < f_D$$

Its impulse response is therefore the inverse Fourier transform

$$h_n(t) = \frac{1}{\sqrt{\pi \cdot f_D}} \cdot \int_{-f_D}^{f_D} \frac{e^{j \cdot 2 \cdot \pi \cdot v \cdot t}}{\left[1 - \left(\frac{v}{f_D}\right)^2\right]^{\frac{1}{4}}} \, dv$$

Remove the singularity by the change of variable $v = f_D \cdot \sin(\theta)$ so that

$$h_n(t) = \sqrt{\frac{f_D}{\pi}} \cdot \int_{-\frac{\pi}{2}}^{\frac{\pi}{2}} \sqrt{\cos(\theta)} \cdot e^{j \cdot 2 \cdot \pi \cdot f_D \cdot t \cdot \sin(\theta)} \, d\theta$$

After taking advantages of symmetries, defining $u = f_D t$ and dropping the constant factors, we have the impulse response as

$$h_n(u) := \int_0^{\frac{\pi}{2}} \sqrt{\cos(\theta)} \cdot \cos(2 \cdot \pi \cdot u \cdot \sin(\theta)) \, d\theta \qquad\qquad TOL \equiv 10^{-5}$$

See what it looks like. Plot one side of it. $u := 0, 0.05 .. 6$

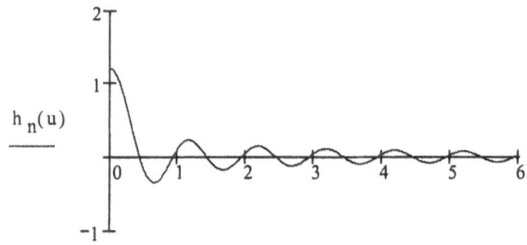

$\overline{h_n(u)}$

It decays *very* slowly, because of the discontinuities in its transform. Zero crossings are just before the integers and half integers.

Calculate the Sampled Impulse Response and Frequency Response

The next step is to sample and truncate the impulse response, then use a Kaiser window to reduce the sidelobes. To start, choose the Doppler frequency f_D as a fraction of the sampling frequency f_s. This gives the normalized sample spacing in time.

Since $\dfrac{f_D}{f_s} = f_D \cdot t_s$ where $t_s = 1/f_s$ is the sample spacing, choose fDts $:= 0.05$

You'll also have to truncate the impulse response to a finite length. The sidelobe level will be lower if you truncate at a zero crossing, so the next several lines help you do this. First, pick a two-sided length expressed in the dimensionless quantity u used as the horizontal axis on the graph above, then find the closest zero crossing to it.

$U := 13$ This is the *two-sided* width; the impulse response support is $[-U/2, U/2]$.

$$U_{xing} := \text{root}\left(h_n\left(\frac{U}{2}\right), U\right) \qquad U_{xing} = 12.876$$

The length of the filter in samples is therefore

$$N_{sb} := \text{floor}\left(\frac{U_{xing}}{fDts}\right) \qquad N_{sb} = 257 \qquad \text{(number of samples per block)}$$

Ideally, this would be a power of 2 for the FFT, but any highly composite number will do. If you don't like the one just above, choose one nearby or choose another value of U.

$N_{sb} := 256$ <=== **and this is where you really select the block size**

Let's capture this as a procedure to help you "negotiate" filter lengths when you use this generation method yourself:

$$\text{filterlength}(U, fDts) := \left| \begin{array}{l} U_{xing} \leftarrow \text{root}\left(h_n\left(\dfrac{U}{2}\right), U\right) \\[2ex] N_{sb} \leftarrow \text{floor}\left(\dfrac{U_{xing}}{fDts}\right) \\[2ex] N_{sb} \end{array} \right.$$

Now we calculate the sampled and truncated impulse response in an array **h**.

$$ib := 0 .. N_{sb} - 1 \qquad h_{ib} := h_n\left[\left[ib - \left(\frac{N_{sb} - 1}{2}\right)\right] \cdot fDts\right]$$

As our final step in preparation of the impulse response, we apply a Kaiser window.

$\beta := 6$ the Kaiser parameter

$$\text{kaiser}_{ib} := \frac{\text{I0}\left[\beta \cdot \sqrt{1 - \left(1 - \frac{2 \cdot ib}{N_{sb} - 1}\right)^2}\right]}{\text{I0}(\beta)}$$ the Kaiser window as an array

The impulse response and the Kaiser window are plotted together below

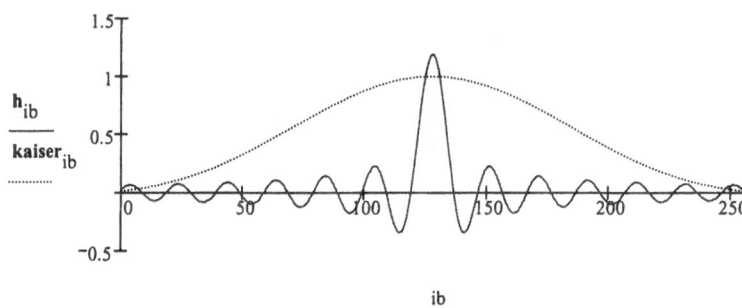

$$\frac{h_{ib}}{\text{kaiser}_{ib}}$$

ib

Now do the windowing $h := (h \cdot \text{kaiser})$

and normalize it to unit sum of squares

$$\text{sumsq} := (h)^2 \qquad h := \frac{h}{\sqrt{\text{sumsq}}} \qquad \text{check:} \qquad \sum_{ib = 0}^{N_{sb} - 1} h_{ib}^2 = 1$$

Finally, the frequency response. We can save some computation by keeping it for use every time. First, a useful array for zero-padding, then the transform:

$$\text{zero}_{ib} := 0$$

$$H := 2 \cdot N_{sb} \cdot \text{CFFT}(\text{stack}(h, \text{zero})) \qquad ib2 := 0 .. 2 \cdot N_{sb} - 1$$

You can use this graph to help you trade off computation (through choice of $f_D t_s$ and N_b) against the quality of the power spectrum approximation:

$$\text{Hc} := \text{recenter}(H) \qquad \text{put the spectrum in the middle of the array for clarity}$$

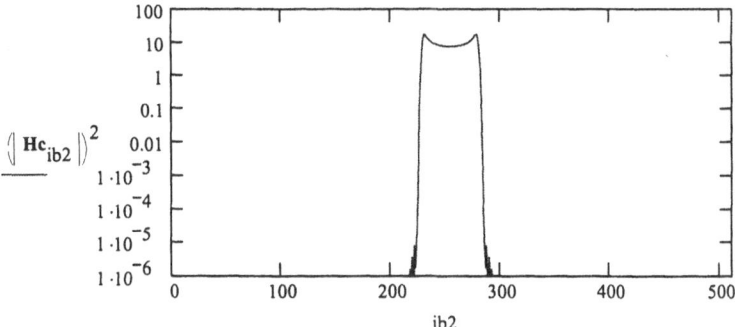

Output Power Spect. (to a scale factor)

Capture the filter generation as a procedure, for more general use:

$$
\text{initfilt}\big(\text{fDts}, N_{sb}, \beta\big) :=
\begin{array}{|l}
\text{for } ib \in 0 .. N_{sb} - 1 \\[1em]
\quad h_{ib} \leftarrow h_n\!\left[\!\left(ib - \left(\dfrac{N_{sb}-1}{2}\right)\right)\!\cdot \text{fDts}\right]\cdot \dfrac{I0\!\left[\beta\cdot\sqrt{1 - \left(1 - \dfrac{2\cdot ib}{N_{sb}-1}\right)^2}\,\right]}{I0(\beta)} \\[2em]
\text{sumsq} \leftarrow (|\,h\,|)^2 \\[1em]
h \leftarrow \text{sumsq}^{-0.5}\cdot h \\[1em]
\text{for } ib \in N_{sb} .. 2\cdot N_{sb} - 1 \\[1em]
\quad h_{ib} \leftarrow 0 \\[1em]
H \leftarrow 2\cdot N_{sb}\cdot \text{CFFT}(h) \\[1em]
H
\end{array}
$$

Overlap-Add Filtering

The overlap-add method [**Oppe89**], in brief, works like this: take a segment of the input signal equal in length to the impulse response; pad the input and the impulse response with zeros to double their lengths; FFT both of them, multiply the transforms, and inverse transform back to the time domain; the second half of the double length convolution is a postcursor, which is saved to be added to the first half of the result the next time the procedure is invoked; the first half of the present array is added to the saved second half from the previous invocation and released as the current result of convolution.

The input signal is a white noise sequence of length equal to the filter length N_{sb}, plus zero padding to twice the filter length. The complex Gaussian generator *cgauss(x)* is explained in **Appendix H**. Note that the variance of the complex gain process will equal 1/2, as recommended in **Section 9.1**.

$$cgauss(x) := \sqrt{-\ln(rnd(1))} \cdot exp(j \cdot rnd(2 \cdot \pi))$$ This ensures that the gain process will have variance equal to 1/2.

$$\mathbf{nshort}_{ib} := cgauss(ib) \qquad \mathbf{n} := stack(\mathbf{nshort}, \mathbf{zero})$$

$$\mathbf{N} := CFFT(\mathbf{n}) \qquad \text{in the frequency domain}$$

We'll keep the current output and the saved second half as columns 0 and 1, respectively, of the array **State**. We initialize it to zero, setting its dimensions in the process, like this:

$$initstate(N_{sb}) := \begin{vmatrix} \text{for } i \in 0..1 \\ \quad \text{for } ib \in 0..N_{sb} - 1 \\ \qquad s_{ib,i} \leftarrow 0 \\ s \end{vmatrix} \qquad \mathbf{State} := initstate(N_{sb})$$

The actual overlap-add filtering step is accomplished, first, by

$$temp := ICFFT\left(\overrightarrow{(\mathbf{N} \cdot \mathbf{H})}\right)$$

which convolves the input and the filter, both of length N_{sb}, and accommodates the tail of the convolution. Next, the output is the sum of the first half of *temp* and the saved tail from last time

$$\mathbf{State}^{<0>} := submatrix(temp, 0, N_{sb} - 1, 0, 0) + \mathbf{State}^{<1>}$$

and we save the second half of *temp* as the tail for the next time around

$$\mathbf{State}^{<1>} := submatrix(temp, N_{sb}, 2 \cdot N_{sb} - 1, 0, 0)$$

Now that we have seen a step-by-step description, it's time to define a complex gain generation procedure for general use

$$filtnoise(\mathbf{State}, \mathbf{H}) := \begin{vmatrix} N_{sb} \leftarrow rows(\mathbf{State}) \\ \text{for } ib2 \in 0..2 \cdot N_{sb} - 1 \\ \quad n_{ib2} \leftarrow if(ib2 < N_{sb}, cgauss(ib), 0) \\ temp \leftarrow \overrightarrow{ICFFT(((CFFT(\mathbf{n}) \cdot \mathbf{H})))} \\ \mathbf{Out}^{<0>} \leftarrow submatrix(temp, 0, N_{sb} - 1, 0, 0) + \mathbf{State}^{<1>} \\ \mathbf{Out}^{<1>} \leftarrow submatrix(temp, N_{sb}, 2 \cdot N_{sb} - 1, 0, 0) \\ \mathbf{Out} \end{vmatrix}$$

Example of Use

All that's left is an example of how to use the procedures to generate two independent complex gain sequences and a demonstration that the result is really continuous from block to block. We'll start by selecting the parameters:

Doppler to sampling ratio	length of generated block	Kaiser parameter

$$fDts := \frac{1}{16} \qquad N_{sb} := 128 \qquad \beta := 6$$

$$H_n := initfilt(fDts, N_{sb}, \beta) \qquad \text{initialize the colouring filter}$$

initialize the states of the two gain processes

$$state1 := initstate(N_{sb}) \qquad\qquad state2 := initstate(N_{sb})$$

discard the startup transients

$$state1 := filtnoise(state1, H_n) \qquad state2 := filtnoise(state2, H_n)$$

Now generate a few blocks of the complex gain processes, storing them in arrays for display

$$state1 := filtnoise(state1, H_n) \qquad\qquad state2 := filtnoise(state2, H_n)$$

$$g1 := state1^{<0>} \qquad\qquad g2 := state2^{<0>}$$

$$state1 := filtnoise(state1, H_n) \qquad\qquad state2 := filtnoise(state2, H_n)$$

$$g1 := stack(g1, state1^{<0>}) \qquad\qquad g2 := stack(g2, state2^{<0>})$$

$$state1 := filtnoise(state1, H_n) \qquad\qquad state2 := filtnoise(state2, H_n)$$

$$g1 := stack(g1, state1^{<0>}) \qquad\qquad g2 := stack(g2, state2^{<0>})$$

And let's have a look at them: $\qquad k := 0 .. rows(g1) - 1$

The continuity at block boundaries is evident from the graph below, as is the independent nature of the processes. Only the real part has been plotted, to reduce the visual clutter, but change it to the imaginary part, if you wish.

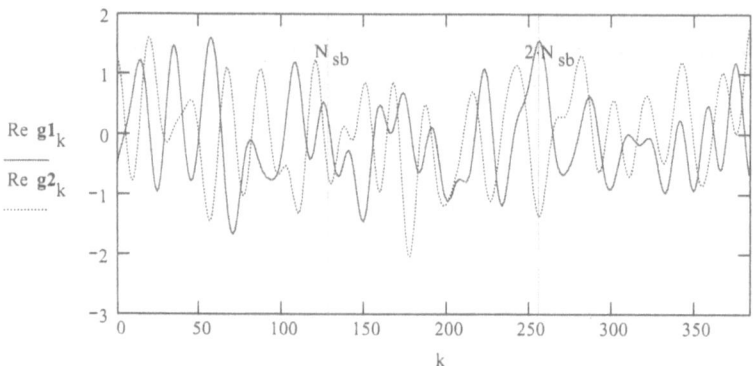

This Appendix has given you four procedures you can use in efficient generation of complex gain sequences of indefinite length:

$h_n(u)$ the ideal filter

filterlength(U, fDts)   to get a suggested value for your filter length

initstate N_{sb} to set up the filter for each gain process

filtnoise(**State**, **H**)   call repeatedly to generate segments of gain process

REFERENCES

Many of the newer references are "active" - a hyperlink takes you to the web site of the author or the author's research group. You can find out about the current work taking place there and, in many cases, download a copy of the referenced work.

[Akai97] Y. Akaiwa, *Introduction to Digital Mobile Communication*, John Wiley and Sons, 1997.

[AlQa97] W.A. Al-Qaq and **J.K. Townsend**, "A Stochastic Importance Sampling Methodology for the Efficient Simulation of Adaptive Ssytems in Frequency Nonselective Rayleigh Fading Channels", *IEEE J. Sel. Areas Commun.*, vol 15, no 4, pp 614-625, May 1997.

[Arno83] H.W. Arnold and W.F. Boldtmann, "Interfade Interval Statistics of a Rayleigh Distributed Wave", *IEEE Trans Commun*, September 1983.

[Bajw82] A.S. Bajwa and **J.D. Parsons**, "Small-area characterisation of UHF urban and suburban mobile radio propagation", *IEE Proc.*, vol. 129, Pt. F, no 2, pp. 102-109, April 1982.

[Beau95] **N.C. Beaulieu**, A.A. Abu-Dayya and **P.J. McLane**, "Estimating the Distribution of a Sum of Independent Lognormal Random Variables", *IEEE Trans Commun*, vol 43, no 12, pp 2869-2873, December 1995.

[Bell63] P.A. Bello, "Characterization of Randomly Time-Variant Linear Channels", *IEEE Trans Commun Syst*, pp 360-393, December 1963.

[Bell64] P.A. Bello, "Time-Frequency Duality", *IEEE Trans Inform Th*, pp 18-33, January 1964.

[Benn96] E. Benner and **A.B. Sesay**, " Effects of Antenna Height, Antenna Gain, and Pattern Downtilting for Cellular Mobile Radio", *IEEE Trans Veh Technol*, vol. 45, no. 2, pp. 217-224, May 1996.

[Bold82] W.F. Boldtmann and H.W. Arnold, "Fade Duration Statistics of Rayleigh Distributed Wave", *IEEE Trans Commun*, March 1982.

[Brit78] British Post Office Standarisation Advisory Group, *A Standard Code for Radiopaging*, 1978.

[Bult87] R.J.C. Bultitude, "Measurement, Characterization and Modelling of Indooor 800/900 MHz Radio Channels for Digital Communications", *IEEE Comms Mag*, vol 25, no 6, pp 5-12, June 1987.

[Bult89a] R.J.C. Bultitude, **S.A. Mahmoud** and W.A. Sullivan, "A Comparison of Indoor Radio Propagation Characteristics at 910 MHz and 1.75 GHz", *IEEE J Selected Areas Commun*, vol 7, no 1, pp 20-30, January 1989.

[Bult89b] R.J.C. Bultitude and G.K. Bedal, "Propagation Characteristics on Microcellular Urban Mobile Radio Channels at 910 MHz", *IEEE J Selected Areas Commun*, vol 7, no 1, pp 31-39, January 1989.

[Butt97] **K.S. Butterworth, K.W. Sowerby and A.G. Williamson**, "Correlated Shadowing in an In-Building Propagation Environment," *Electronics Letters*, Vol. 33, No 5, pp. 420-421, February 1997.

[Butt98] **K. S. Butterworth, K. W. Sowerby, A. G. Williamson, and M. J. Neve**, "Influence of In-Building Correlated Shadowing for Different Buildings and Base Station Configurations on System Capacity," *Proc. 48th IEEE Veh. Technol. Conf.*, pp. 850-855, Ottawa, Canada, May 1998.

[Cave92] **J.K. Cavers** and **P. Ho**, "Reducing the Computation Time in Simulations of Fading Channels", *IEEE Veh Technol Conf*, Denver, 1992.

[Cave95] **J.K. Cavers**, "Pilot Symbol Assisted Modulation and Differential Detection in Fading and Delay Spread", *IEEE Trans Veh Technol*, vol 43, no 7, pp 2206-2212, July 1995.

[Chia90] S.T.S. Chia, "Radiowave Propagation and Handover Criteria for Microcells", *British Telecom Technol J*, vol 8, no 4, pp 50-61, October 1990.

[Cimi85] **L.J. Cimini**, "Analysis and Simulation of a Digital Mobile Channel Using Orthogonal Frequency Division Multiplexing", *IEEE Trans Commun*, vol 33, no 7, pp 665-675, July 1985.

[Couc87] **L.W. Couch**, *Digital and Analog Communication Systems*, 2nd ed., MacMillan, 1987.

[Dari92] **G. D'Aria, F. Muratore and V. Palestini**, "Simulation and Performance of the Pan-European Land Mobile Radio System", *IEEE Trans Veh Technol*, vol 41, no 2, pp 177-189, May 1992.

[Dave87] W.B. Davenport and W.L. Root, *An Introduction to the Theory of Random Signals and Noise*, McGraw-Hill 1958 and IEEE Press 1987 .

[Deng95] G. Deng, **J. Cavers** and **P. Ho**, "A Reduced Dimensionality Propagation Model for Frequency Selective Rayleigh Fading Channels", *IEEE Intl Conf on Commun*, Seattle, 1995.

[Dent93] P. Dent, G.E. Bottomley and T. Croft, "Jakes Fading Model Revisited", *IEE Electronics Letters*, vol. 29, no.3, pp. 1162-1163, June 1993.

[Erce99] V. Erceg, **D.G. Michelson, S.S. Ghassemzadeh, L.J. Greenstein, A.J.Rustako, Jr.,** P.B. Guerlain, M.K. Dennison, **R.S. Roman**, D.J. Barnickel, S.C. Wang, and **R.R. Miller**, "A model for the multipath delay profile over narrowband fixed wireless channels," *IEEE J. Sel. Areas Communic.*, vol. 17, no. 3, pp. 399-410, Mar. 1999.

[Fech93] **S.A. Fechtel**, "A Novel Approach to Modeling and Efficient Simulation of Frequency- Selective Fading Radio Channels", *IEEE J Selected Areas in Commun*, vol 11, no 3, pp 422-431, April 1993.

[Fren79] R.C. French, "The Effect of Fading and Shadowing on Channel Reuse in Mobile Radio", *IEEE Trans Veh Technol*, vol VT-28, no 3, pp 171-181, August 1979.

[Gree90] E. Green, "Radio Link Design in Microcellular Systems", *British Telecom Technology J*, vol. 8, no. 1, pp. 85-96, Jan. 1990.

[Hash79] **H. Hashemi**, "Simulation of the Urban Radio Propagation Channel", *IEEE Trans Veh Techn*, vol VT-28, n0 3, pp 213-225, August 1979.

[Hash93a] **H. Hashemi**, "The Indoor Radio Propagation Channel", *Proc. IEEE*, vol 81, no 7, pp 943-968, July 1993.

[Hash93b] **H. Hashemi**, "Impulse Response Modelling of Indoor Radio Propagation Channels", *IEEE J Select Areas in Commun*, vol 11, no 7, pp 967-978, September 1993.

[Hata80] M. Hata, "Empirical Formula for Propagation Loss in Land Mobile Radio Services", *IEEE Trans Veh Technol*, vol. 29, no. 3, pp. 317-325, Aug. 1980.

[Hayk96] S. Haykin, *Adaptive Filter Theory*, 3rd ed., Prentice-Hall, 1996.

[IEEE88] Special Issue on Mobile Radio Propagation, *IEEE Trans Veh Technol*, vol. 37, no. 1, Feb. 1988. See also **Propagation Committee**.

[Jake74] W.C. Jakes, ed., *Microwave Mobile Communications*, AT&T, 1974, reissued by IEEE, 1993.

[Jans96] **G.J.M. Janssen**, P.A. Stigter and **R. Prasad**, "Wideband Indoor Channel Measurements and BER Analysis of a Frequency Selective Multipath Channels at 2.4, 4.75 and 11.5 GHz", *IEEE Trans Commun*, vol 44, no 10, pp 1272-1288, October 1996.

[Jeru92] M.C. Jeruchim, P. Balaban and **K.S. Shanmugan**, *Simulation of Communication Systems*, Plenum Press, 1992.

[Kalk97] M. Kalkan and R.H. Clarke, "Prediction of the Space-Frequency Correlation Function for Base Station Diversity Reception", *IEEE Trans. Veh. Technol.*, vol. 46, no. 1, pp. 176-184, February 1997.

[Keen90] J.M. Keenan and A.J. Motley, "Radio Coverage in Buildings", *British Telecom Technol J*, vol 8, no 1, pp 19-24, January 1990.

[Kenn69] R.S. Kennedy, *Fading Dispersive Communication Channels*, John Wiley, 1969. POCSAG

[Korn90] I. Korn, "M-ary Frequency Shift Keying with Limiter-Discriminator-Integrator Detector in Satellite Mobile Channel with Narrow-Band Receiver Filter", *IEEE Trans Commun*, vol 38, no 10, pp 1771-1778, October 1990.

[Lee82] W.C.Y. Lee, *Mobile Communications Engineering*, McGraw-Hill, 1982.

[Leta97] **K. ben Letaief**, K Muhammad and J.S. Sadowsky, "Fast Simulation of DS/CDMA With and Without Coding in Multipath Fading Channels", *IEEE J. Sel. Areas Commun.*, vol 15, no 4, pp 626-639, May 1997.

[Mich99] **D.G. Michelson**, V. Erceg, and **L.J. Greenstein**, "Modeling diversity reception over narrowband fixed wireless channels," *IEEE MTT-S Int. Topical Symp. Tech. Wireless Applic.* (Vancouver, BC), 21-24 Feb. 1999, pp. 95-100.

[Oppe89] A.V. Oppenheim and R.W. Schafer, *Discrete-Time Signal Processing*, Prentice-Hall, 1989.

[Orfa85] S.J. Orfanidis, *Optimum Signal Processing: An Introduction*, Macmillan, 1985.

[Naka60] N. Nakagami, "The m-distribution, a general formula for intensity distribution of rapid fading", in *Statistical Methods in Radio Wave Propagation*, W. Hoffman, ed., Pergamon, 1960.

[Panj96] M. A. Panjawani, A. L. Abbott, **T. S. Rappaport**, "Interactive Computation of Coverage Regions for Wireless Communication in Multifloored Indoor Environments," *IEEE Journal on Selected Areas in Communications*, vol. 14, no. 3, pp. 420-430, March 1996.

[Pars89] **J.D. Parsons** and J.G. Gardiner, *Mobile Communication Systems*, Blackie and Son, London, 1989.

[Papo84] A. Papoulis, *Probability, Random Variables and Stochastic Processes*, 2nd ed, McGraw-Hill, 1984.

[Pate97] **F. Patenaude, J.H. Lodge** and **J.-Y. Chouinard**, "Error probability expressions for non-coherent diversity in Nakagami fading channels", *IEEE Veh Technol Conf*, Phoenix, 1997.

[Proa89] **J.G. Proakis**, *Digital Communications*, 2nd edition, McGraw-Hill, 1989.

[Proa95] **J.G. Proakis**, *Digital Communications*, 3rd edition, McGraw-Hill, 1995.

[Rice44] S.O. Rice, "Mathematical Analysis of Random Noise", Part 1, *Bell Syst Tech J*, vol 23, pp 282-332, July 1944.

[Rice45] S.O. Rice, "Mathematical Analysis of Random Noise", Part 2, *Bell Syst Tech J*, vol 24, pp 46-156, January 1945.

[Sado98] J.S. Sadowsky and V. Kafedziski, "On the Correlation and Scattering Functions of the WSSUS Channel for Mobile Communications", *IEEE Trans Veh Technol*, vol 47, no 1, pp 270-282, February 1998.

[Salt67] B.R. Salzburg, "Performance of an Efficient Parallel Data Transmission System", *IEEE Trans Commun Techn*, vol COM-15, no 6, pp 805-811, December 1967.

[Salz94] J. Salz and **J.H. Winters**, "Effect of Fading Correlation on Adaptive Arrays in Digital Mobile Radio", *IEEE Trans Veh Technol*, vol 43, no. 4, November 1994.

[Schw82] S.C. Schwartz and Y.-S. Yeh, "On the distribution function and moments of power sums with lognormal components", *Bell Syst Tech J*, vol 61, pp 1441-1462, September 1982.

[Seid94] S. Y. Seidel, **T. S. Rappaport**, "Site-Specific Propagation Prediction for Wireless In-Building Personal Communication System Design," IEEE Trans. Veh. Technol., vol. 43, no. 4, pp. 879-891, November 1994.

[Smit97] **P.J. Smith**, M. Shafi and H. Gao, "Quick Simulation: A Review of Importance Sampling Techniques in Communications Systems", *IEEE J. Sel. Areas Commun.*, vol 15, no 4, pp 597-613, May 1997.

[Sous95] **E.S. Sousa**, "Antenna Architectures for CDMA Integrated Wireless Access Networks", *Proc IEEE PIMRC*, pp 921-925, Toronto, September 1995.

[Sous94] **E.S. Sousa**, V.M. Jovanovic and C. Deigneault, "Delay Spread Measurements for the Digital Cellular Channel in Toronto", *IEEE Trans Veh Technol*, vol 43, no 4, pp 837-847, November 1994.

[Stee92] **R. Steele**, *Mobile Radio Communications*, Pentech Press, 1992.

[Stub96] **G.L. Stuber**, *Principles of Mobile Communication*, Kluwer Academic Press, 1996.

[Taga99] T. Taga, T. Furuno and K. Suwa, "Channel Modeling for 2-GHz-Band Urban Line-of-Sight Street Microcells", *IEEE Trans Veh Technol*, vol 48, no.1, pp. 262-272, January 1999.

[Taub86] H. Taub and D.L. Schilling, *Principles of Communication Systems*, McGraw-Hill, 1987.

[Todd92] S.R. Todd, **M.S. El-Tanany** and **S.A. Mahmoud**, "Space and Frequency Diversity Measurements of the 1.7 GHz Indoor Radio Channel Using a Four-Branch Receiver", *IEEE Trans Veh Technol*, vol 41, no 3, August 1992.

[Ugwe97] O.C. Ugweje and **V.A. Aalo**, "Performance of selection diversity system in correlated Nakagami fading", *IEEE Veh Technol Conf*, Phoenix, 1997.

[Vija93] R. Vijayan and J.M. Holtzman, "A Model for Analyzing Handoff Algorithms", *IEEE Trans Veh Technol*, vol 42, no 3, pp 351-356, August 1993.

[Wu94] G. Wu, A. Jalali and **P. Mermelstein**, "On Channel Models for Microcellular CDMA", *IEEE Veh Techn Conf*, 1994.

[Yaco93] M.D. Yacoub, *Foundations of Mobile Radio Engineering*, CRC Press, 1993.

[Yeh84] Y.-S. Yeh and S.C. Schwartz, "Outage Probability in Mobile Telephony Due to Multiple Log-Normal Interferers", *IEEE Trans Commun*, vol COM-32, no 4, pp 380-388, April 1984.

[Ziem76] R.E. Ziemer and **W.H. Tranter**, *Principles of Communications*, Houghton Mifflin, 1976.

INDEX OF TOPICS

Adjacent channel interference (ACI) **1.2**
Angle of arrival **3.2**, **5.1**, **5.2**, **5.3**, **8.3**
Angular dispersion **8.3**
Angular power density **8.3**
Antenna
    arrays **8.3**
    beamwidth **8.1**
    directionality at base station **8.3**
    directionality at mobile **5.1**, **8.1**, **8.2**
    elevation **1.1**, **7.1**, **7.2**
    gain **5.1**, **8.1**, **8.3**
Autocorrelation function of the channel complex gain
    at mobile **Appendix F**, **5.1**, **8.1**
    at base station **8.3**
BER analysis
    AWGN channel **4.3**
    Rayleigh fading channel **4.3**, **6.1**, **6.2**
BER simulation **9.1**, **9.3**
Binomial coefficient **Unit and function definitions**
Carrier-to-interference ratio (C/I) **1.2**, **2.2**
Cellular clusters **1.2**
Channel complex gain **3.2**
    animation **Appendix D**
    Gaussian approximation **Appendix J**
    generation for importance sampling **Appendix H**
    generation using filtered white noise **9.2**
    generation using Jake's method **Appendix B**, **3.1**, **9.1**, **9.2**, **9.3**
    joint second order statistics **Appendix F**, **Appendix G**
    scaling for simulations **9.1**
Channel frequency response **3.3**
    animation **Appendix E**
Channel impulse response **3.3**
    animation **Appendix E**
Channel models
    mathematical **3.2**, **3.3**
    physical **3.1**
    tapped delay line (TDL) **5.4**, **9.1**
    hilly, rural, and urban environments **7.1**
Chi-square distribution **4.2**
Cochannel cells **1.2**
Cochannel interference (CCI) **1.2**, **2.2**

Coherence
    bandwidth **5.2**
    distance **8.1**, **8.3**
    time **5.1**, **8.1**
Complex Gaussian random vectors **Appendix F**
Coverage holes **2.2**
Delay bins (see Channel models, tapped delay line (TDL))
Delay moments **5.2**
Delay-power profile, see Power-delay profile
Delay spread **3.3**, **5.2**
    effect of antenna directionality **8.2**
Delay-Doppler spread function (see Scattering function)
Discriminator, see Limiter-discriminator detection
Diversity reception **7.2**, **8.1**, **8.3**
Doppler shift **3.2**, **5.1**
Doppler spectrum **5.1**, **5.3**, **8.1**
Doppler spread **3.3**, **5.1**
    effect of antenna directionality **5.1**, **8.1**
Error bursts **6.2**, **4.3**
Error floor **3.3**, **6.1**
Fading
    physical basis **3.1**
    mathematical model **3.2**
    Rayleigh **4.2**
    Rice **4.2**, **7.1**, **7.2**, **7.3**
    fast, slow **3.3**
    flat **3.3**, **3.4**, **4.4**, **5.1**
    frequency selective **3.3**, **3.4**, **4.4**, **5.3**, **5.4**
    statistics **6.1**, **6.2**, **7.1**, **7.2**, **7.3**
    frequency and duration of fades **6.2**
    inter-fade duration **6.2**
Frequency offset **8.1**
Frequency shift keying (FSK) **6.1**
Gaussian channel **4.1**, **Appendix J**
Handoff **2.2**
Handover, see Handoff
Hexagonal cells **1.2**
Importance sampling **9.3**, **Appendix H**
Interference-limited operation **1.2**
Intersymbol interference (ISI) **3.3**
Irreducible error rate, see Error floor
Isotropic scattering **5.1**, **Appendix B**
Level crossing rate **6.2**, **8.1**
Limiter, see Limiter-discriminator detection
Limiter-discriminator detection **6.1**

Line of sight (LOS) signal **3.2**, **5.1**, **5.2**, **5.3**
Linear time-invariant (LTI) systems **Appendix I**
Linear time-variant (LTV) systems **Appendix I**
Lognormal distribution **7.1**, **Appendix A**
Macrocells **7.1**
Microcells, urban **7.2**
Nakagami distribution **4.2**, **7.1**
Near-far effect **1.2**
Noise, sampling and scaling for simulations **9.1**
Noise-limited operation **1.2**
Path Loss
    implications for cellular design **1.2**
    models **1.1**, **3.1**, **4.4**
Path loss exponent **1.1**
    indoor picocells **7.3**
    macrocells **7.1**
    urban microcells **7.2**
Phase-derivative, PDF of, **6.1**, **Appendix G**
Phase-lock loop (PLL) detection **6.1**
Picocells, indoor **7.3**
Power delay profile **5.2**, **5.3**
    exponential **7.1**, **7.2**
    indoor picocells **7.3**
    macrocells **7.1**
    urban microcells **7.2**
Power spectrum of the complex gain **5.1**, **Appendix F**
Propagation exponent, see Path loss exponent
Q-function **Unit and function definitions**
Quasi-synchronous transmission **2.2**
Random FM **6.1**, **Appendix D**, **Appendix G**
Random standing wave, visualization of, **Appendix C**
Rayleigh distribution **4.2**, **7.1**
Reflection coefficient **3.2**
Rice distribution **4.2**, **7.1**, **7.2**
Rice K Factor
    indoor picocells **7.3**
    macrocells **7.1**
    urban microcells **7.2**
RMS delay spread, **5.2**
    indoor picocells **7.3**
    macrocells **7.1**
    urban microcells **7.2**
RMS Doppler spread **5.1**
Scattering function **5.3**, **8.2**

Shadowing
  lognormal **2.1**, **4.4**, **7.1**, **7.3**
  physical basis of, **2.0**, **3.1**
  safety margin **2.2**
  statistical model **2.1**
  in system design **2.2**
Signal-to-noise-ratio (SNR) **4.3**, **9.1**
Simulcast transmission **2.2**
Simulation of fading channels **9.1**, **9.2**, **9.3**
Spaced frequency correlation function **5.2**, **7.1**
  typical channels **7.1**
Static channel, frequency selective **5.2**
Suzuki distribution **4.4**
System performance in fading **6.1**, **6.2**
Tapped delay line (TDL) model, see Channel, TDL
Time-frequency correlation function **5.3**
Time-variant filters **3.3**, **Appendix I**
Uncorrelated scattering (US) **5.2**, **5.3**
Unit and function definitions
Wide sense stationary (WSS) **5.1**, **5.3**
Wide sense stationary uncorrelated scattering (WSSUS) **5.3**

About the CD-ROM

Mobile Channel Characteristics was conceived and written as an interactive text to be viewed on a computer screen. It includes many features not found in conventional texts:

- The entire text resides on your hard drive. It is always ready, just a mouse-click away.
- It is a live document. Try different parameter values, and the equations, tables and graphs recalculate as you watch. Animated graphs illustrate dynamics of the channel. Explore propagation, modulation or system models interactively to gain additional insight.
- The examples and appendices are "tear-off design sheets". You can use their programs on the job or in your thesis to speed up your work.
- It links you to the world. Hyperlinks connect you to websites of cited authors, to online research journals and to employers and graduate schools, all through the Internet.

Installing with Windows 95/98 and Windows NT 4.0

Insert the CD-ROM in your CD-ROM drive. The installation dialog begins automatically and you will be guided through the installation. *Mobile Channel Characteristics* occupies 9.5 Mbyte of hard drive space.

You must have version 8 or higher of Mathsoft's Mathcad or Mathcad Explorer on your computer **at the time you install this text**. If you do not have either of them, the installation dialog helps you download a free copy of Mathcad Explorer from Mathsoft (www.mathsoft.com), **provided you are connected to the Internet.** The download file is 12.4 Mbyte. It is recommended that you run Mathcad Explorer once immediately after its installation before the first time you open *Mobile Channel Characteristics*.

One section of *Mobile Channel Characteristics* contains images which can be viewed more clearly with Adobe Acrobat Reader. This is optional, not essential. If you do not have it, the installation dialog allows you to install Adobe Acrobat Reader 4.05 directly from the CD-ROM.

Disclaimers

This CD-ROM is distributed by Kluwer Academic Publishers with *ABSOLUTELY NO SUPPORT* and * NO WARRANTY * from Kluwer Academic Publishers.

Kluwer Academic Publishers shall not be liable for damages in connection with, or arising out of, the furnishing, performance or use of this CD-ROM.